Novel Advances in Aquatic Vegetation Monitoring in Ocean, Lakes and Rivers

Novel Advances in Aquatic Vegetation Monitoring in Ocean, Lakes and Rivers

Special Issue Editor

Monica Rivas Casado

MDPI • Basel • Beijing • Wuhan • Barcelona • Belgrade

MDPI

Special Issue Editor
Monica Rivas Casado
Cranfield University
UK

Editorial Office
MDPI
St. Alban-Anlage 66
4052 Basel, Switzerland

This is a reprint of articles from the Special Issue published online in the open access journal *Remote Sensing* (ISSN 2072-4292) from 2018 to 2019 (available at: https://www.mdpi.com/journal/remotesensing/special_issues/aquatic)

For citation purposes, cite each article independently as indicated on the article page online and as indicated below:

LastName, A.A.; LastName, B.B.; LastName, C.C. Article Title. *Journal Name* **Year**, *Article Number*, Page Range.

ISBN 978-3-03921-205-7 (Pbk)
ISBN 978-3-03921-206-4 (PDF)

Contents

About the Special Issue Editor

Monica Rivas Casado is a Senior Lecturer in Integrated Environmental Monitoring with expertise in the application of statistics to environmental data. Her academic career has been built around the integration of emerging technologies, advanced statistics, and environmental engineering for the design of robust monitoring strategies. She has an MSc in Environmental Water Management and a PhD in applied geostatistics from Cranfield. She is a Chartered Environmentalist (CEnv), a Chartered Scientist (CSci), a Chartered Forestry Engineer, and a Fellow of the Higher Education Academy (FHEA). She holds a Postgraduate Certificate (PGCert) in Teaching and Learning and a National Examination Board in Occupational Safety and Health (NEBOSH) qualification. Monica is a fully qualified RPQ-s (drone) pilot. Monica is currently leading Research Council (RCUK)- and industry-funded projects on the use of emerging technologies and statistical science for robust environmental monitoring. This includes the use of drones for jellyfish and seaweed bloom monitoring near coastal nuclear power plants, drones for flood and catastrophe extent mapping and damage assessment, autonomous on-water vehicles for robust hydromorphological characterization, and floodplain forest restoration monitoring. Further research activity includes underwater vehicles for coral reef habitat identification and ecosystem service quantification. She is the lead at Cranfield University for the Building Resilience Into Risk Management (BRIM) EPSRC Grand-Challenge network (2015–2019). Previously, she has lead research activity on geostatistical science for the design of monitoring programs for mycotoxins and successfully contributed and delivered RCUK research on ecosystem services science. She has collaborated with a vast array of user organizations on: (i) the design of biodiversity monitoring programs within Integrated Biodiversity Delivery Areas, (ii) the quantification of the effect of forest management practices on soil carbon sequestration, and (iii) the assessment of land use dynamics in Europe, amongst others. Her work has been used as a reference for best practice guideline development and policy implementation. During her 10 month secondment with a governmental department (Department for Business Innovation and Skills, Government for Science, 2014–2015), she provided advice and guidance on emerging technologies to different ministerial departments, including the Cabinet Office.

Preface to "Novel Advances in Aquatic Vegetation Monitoring in Ocean, Lakes and Rivers"

In recent decades, there has been an increase in the development of strategies for water ecosystem mapping and monitoring. Overall, this is primarily due to legislative efforts to improve the quality of water bodies and oceans. Remote sensing has played a key role in the development of such approaches—from the use of drones for vegetation mapping to autonomous vessels for water quality monitoring. Within the specific context of vegetation characterization, the wide range of available observations—from satellite imagery to high-resolution drone aerial imagery—has enabled the development of monitoring and mapping strategies at multiple scales (e.g., micro- and mesoscales).

This Special Issue collates recent advances in remote sensing-based methods applied to ocean, river, and lake vegetation characterization, including seaweed/kelp, submerged and emergent vegetation, and floating-leaf and free-floating plants. A total of six manuscripts have been compiled in this Special Issue and a brief description of each paper is provided below.

In "Decision-Tree, Rule-Based, and Random Forest Classification of High-Resolution Multispectral Imagery for Wetland Mapping and Inventory", the authors explore multiple machine learning algorithms for classifying wetland-dominated landscapes. The author's primary conclusion was that the Random Forest should be the classifier of choice in most cases.

The authors of "A New Method for Mapping Aquatic Vegetation Especially Underwater Vegetation in Lake Ulansuhai Using GF-1 Sattelite Data" proposed a new concave–convex decision function to detect submerged aquatic vegetation and identify bodies of water using Gao Fen multispectral satellite images. Their contribution shows that the proposed decision function outperformed traditional classification methods in distinguishing water and submerged aquatic vegetation.

In "Using 1st Derivative Reflectance Signatures within a Remote Sensing Framework to Identify Macroalgae in Marine Environments", the work presented focuses on the impacts of macroalgae blooms on the operations of British nuclear power stations. The authors analyze the spectral reflectance properties of the problematic macroalgae species. The authors use their results to inform the development of a drone-based early warning system for macroalgae detection.

In the article "Performance Evaluation of Newly Proposed Seaweed Enhancing Index", spectral bands of near-infrared and shortwave-infrared Landsat 8 satellite data are used to develop a new remote sensing-based seaweed enhancing index. The authors report the enhanced performance of the developed index when compared to Normalized Difference Vegetation Index (NDVI)-derived results.

The authors of "Rapid Invasion of Spartina alterniflora in Coastal Zone of Mainland China: New Observations from Landsat OLI Images" explore the use of object-based image analysis and support vector machine methods to better understand the spatial variability of S. alterniflora in coastal areas. The authors claim in their contribution that monitoring methods enabling geospatially varied responding decisions are needed to promote sustainable coastal ecosystems.

In "Mapping Substrate Types and Compositions in Shallow Streams", the impacts of water column correction for substrate mapping in shallow fluvial systems is investigated. The authors report on how the red-edge band of WV3 considerably improves the characterization of submerged aquatic vegetation densities from both above-water and retrieved bottom spectra.

This Special Issues compiles a range of novel contributions that will be of interest to all readers interested in "Novel Advances in Aquatic Vegetation Monitoring in Ocean, Lakes and Rivers". We thank all authors and co-authors for their thoughtful contributions and hope the work presented inspires further research within this field.

Monica Rivas Casado
Special Issue Editor

remote sensing

MDPI

Article

Decision-Tree, Rule-Based, and Random Forest Classification of High-Resolution Multispectral Imagery for Wetland Mapping and Inventory

Tedros M. Berhane [1], Charles R. Lane [2,*], Qiusheng Wu [3], Bradley C. Autrey [2],
Oleg A. Anenkhonov [4], Victor V. Chepinoga [5,6] and Hongxing Liu [7]

[1] Pegasus Technical Services, Inc., C/O U.S. Environmental Protection Agency, Cincinnati, OH 45219, USA;
     berhane.tedros@epa.gov
[2] Office of Research and Development, U.S. Environmental Protection Agency, Cincinnati, OH 45268, USA;
     autrey.brad@epa.gov
[3] Department of Geography, Binghamton University, State University of New York,
     Binghamton, NY 13902, USA; wqs@binghamton.edu
[4] Laboratory of Floristics and Geobotany, Institute of General and Experimental Biology SB RAS,
     670047 Ulan-Ude, Russia; anen@yandex.ru
[5] Laboratory of Physical Geography and Biogeography, V.B. Sochava Institute of Geography SB RAS,
     664033 Irkutsk, Russia; victor.chepinoga@gmail.com
[6] Department of Botany, Irkutsk State University, 664003 Irkutsk, Russia
[7] Department of Geography, University of Cincinnati, Cincinnati, OH 45220, USA; hongxing.liu@uc.edu
*   Correspondence: lane.charles@epa.gov; Tel.: +1-513-569-7854

Received: 7 March 2018; Accepted: 2 April 2018; Published: 9 April 2018

Abstract: Efforts are increasingly being made to classify the world's wetland resources, an important ecosystem and habitat that is diminishing in abundance. There are multiple remote sensing classification methods, including a suite of nonparametric classifiers such as decision-tree (DT), rule-based (RB), and random forest (RF). High-resolution satellite imagery can provide more specificity to the classified end product, and ancillary data layers such as the Normalized Difference Vegetation Index, and hydrogeomorphic layers such as distance-to-a-stream can be coupled to improve overall accuracy (OA) in wetland studies. In this paper, we contrast three nonparametric machine-learning algorithms (DT, RB, and RF) using a large field-based dataset ($n = 228$) from the Selenga River Delta of Lake Baikal, Russia. We also explore the use of ancillary data layers selected to improve OA, with a goal of providing end users with a recommended classifier to use and the most parsimonious suite of input parameters for classifying wetland-dominated landscapes. Though all classifiers appeared suitable, the RF classification outperformed both the DT and RB methods, achieving OA >81%. Including a texture metric (homogeneity) substantially improved the classification OA. However, including vegetation/soil/water metrics (based on WorldView-2 band combinations), hydrogeomorphic data layers, and elevation data layers to increase the descriptive content of the input parameters surprisingly did not markedly improve the OA. We conclude that, in most cases, RF should be the classifier of choice. The potential exception to this recommendation is under the circumstance where the end user requires narrative rules to best manage his or her resource. Though not useful in this study, continuously increasing satellite imagery resolution and band availability suggests the inclusion of ancillary contextual data layers such as soil metrics or elevation data, the granularity of which may define its utility in subsequent wetland classifications.

Keywords: freshwater wetland; Lake Baikal; methodological comparison; Selenga River Delta; WorldView-2

1. Introduction

Wetlands are dynamic environments existing at the terrestrial-aquatic interface. As such, they are vulnerable to a wide range of human-mediated environmental and hydrological alterations associated with population growth, urbanization, and increased human development activities. Global and regional climate change, particularly temperature alterations and changing precipitation trends, have considerably affected wetland ecosystems [1,2]. Despite their vital functions in biodiversity and ecosystem services (e.g., [3]), wetlands have experienced extensive losses throughout the world in recent decades [4–8]. Intelligent planning measures and effective management policies need to be formulated to conserve and protect existing wetland resources; to mitigate negative anthropogenic impacts on wetlands; and to maintain wetland integrity, functioning, and resilience. Wetland mapping and inventory are critical to acquiring the scientific knowledge about wetland habitats, including their location, extent, and spatial distribution, as well as their vegetation composition, structure, and density. Once this knowledge is gained, effective management can ensue.

Satellite data have long been utilized to augment and supplant field- and aerial-based assessment techniques [9,10]. However, due to their high spatial heterogeneity and temporal hydrologic variability, wetlands have been among the most difficult ecosystems to classify with remotely sensed imagery [9,11–14]. In past decades, automated image classification approaches were extensively adopted to process satellite remote sensing imagery for mapping and studying wetlands at a large spatial scale, reducing inconsistencies associated with human interpretation, and creating reproducible wetland maps [15–17]. Satellite data require unsupervised or supervised classification of the spectral signatures for wetland characterization, and classification approaches have advanced in concert with satellite advancements [9,18–27]. There is now a wealth of classifiers, including Iterative Self-Organizing Data Analysis Technique (ISODATA) [28–30], maximum likelihood [30,31], artificial neural network [32,33], support vector machine [34], and ensemble approaches [35].

Three increasingly employed approaches for classifying remote sensing images are decision-tree (DT), rule-based (RB), and random forest (RF) classification. All three methods are nonparametric, and as such they are not constrained by the distribution of the predictor variables. The DT method is an efficient inductive machine learning technique [36–38]. A DT consists of a root-nodes-branches-leaf flowchart that is created to effectively bin data by recursively testing attributes of the dataset at each tree node, with branches representing the different outcomes leading to subsequent nodes, until a leaf (or terminal node) is created, representing a class. Compared with traditional classification methods such as the maximum likelihood and linear discriminant function classifiers, the DT method has a number of advantages [39,40]. As a nonparametric classifier, it is robust with respect to nonlinear interactions between variables and relatively insensitive to noisy relationships between input attributes and class labels [41]. It makes no assumptions regarding normality for the predictor variables and can easily accommodate both continuous and categorical data from various measurement scales (i.e., nominal, ordinal, interval, and ratio scales [36–38]). Examples of wetland classifications using DTs from the literature include Baker et al. [42], who used a DT-based classification method with Landsat Enhanced Thematic Mapper Plus imagery and both topographic and soil data to distinguish wetlands and riparian areas with 86% overall accuracy. Similarly, Wright and Gallant [13] used DTs to combine Landsat Thematic Mapper imagery and ancillary environmental data to discriminate among five palustrine wetland types in a large national park in the western US.

The RB approach creates a series of "if-then" rules to effectively classify landscapes, and can similarly couple different types of data in the process (e.g., [39,43]). The RB approach is similar to the DT approach, but generally has fewer rules and contains contextual information within the ruleset, hence it is simpler to understand than the complex bifurcating DTs. Domain knowledge, spatial context, and associations can also be integrated into the RB classification algorithm. For instance, Li and Chen [44] used Landsat ETM+, Radarsat Synthetic Aperture Radar (SAR), and elevation data in a series of "if-then" rules to classify each pixel in their study area as open bog, open fen, tree bog, marsh, or swamp, with classification accuracy ranging from 71 to 92%. Sader et al. [45] applied RB methods to effectively classify Landsat TM images for discrimination between forested wetlands and uplands. Houhoulis and Michener [46] created an RB approach to detect wetland change coupling SPOT-XS imagery and aerially derived wetland inventory data.

The RF approach is a relatively novel classification technique based on ensemble machine learning and has been increasingly used as a classifier of choice for remote sensing of different types of wetlands and aquatic habitats (e.g., [47]). The RF approach, like DT and RB, is nonparametric, robust to normal distribution departures, and can be used for both classifications and regressions, as well as determining variable importance [48,49]. Thus, RF has many of the benefits of DT and RB classification while overcoming several limitations, such as overfitting [50]. Errors of bias and variance can affect the DT and RB approaches, whereas RF avoids both errors through random selection of input-predictor variables and use of different subsets of the same training dataset [50,51]. Examples of RF used in wetland classification include Tian et al. [52], who used fused Pléiade-1B and multitemporal Landsat-8 data for mapping wetland cover in an arid region in China. Corcoran et al. [53] used multisource and multitemporal remote sensing and ancillary information such as radar and optical data, topographic data, and soil characteristics for mapping mixed managed and natural wetlands of woody and herbaceous plants in Minnesota, USA. Van Beijma et al. [54] used both the S-band and X-band quad-polarimetric airborne SAR, elevation data, and optical remotely sensed data for mapping natural coastal salt marsh vegetation habitats. Disadvantages of RF include a relatively longer processing time and model complexity, especially in comparison with the DT and RB methods [51].

Increased interest in wetland ecosystems has resulted in marked advances in characterizing and classifying wetland structure (frequently controlled by inundation patterning; see [55,56]). An abundance of advanced satellite platforms and spatial data coupled with an understanding of differences in functional rates based on wetland typologies (e.g., [57,58]) portends a need to choose effective classification techniques. Which technique is chosen will hinge on the needs of the end user, who could consider trade-offs such as the simplicity of DT or contextual nature of RB versus the classification robustness of the higher-complexity RF approach. In this paper, we contrast and report the efficacy and accuracy of three different classifiers, the DT, RB, and RF approaches, using a large wetland vegetation dataset and high-resolution imagery. As end users might have different resources available, we further explore changes in overall accuracy (OA) using parsimonious inputs (e.g., a spartan four-band analysis) as well as highly parameterized inputs (e.g., eight bands plus spectral metrics, hydrogeomorphic variables, and elevation data). The goal of this study is to provide methodological recommendations to effectively, efficiently, and robustly classify a given wetland landscape.

2. Methods

2.1. Study Area

The study area includes the Kabansky Nature Reserve and the surrounding area within the ~600 km^2 Selenga River Delta in southeastern Siberia, Russia [59]. The delta has a variety of wetland habitats, from open water to emergent marshes, shrub scrub, forested wetlands, and mixed habitats [60,61]. The delta traps suspended sediments and filters excess nutrients, heavy metals, and other pollutants from the Selenga River [59,62]. The Selenga River Delta was defined as a Ramsar Wetland of International Importance in 1994 for its significant flora and fauna [63]. The Selenga River, which provides sediment and water resources to maintain the delta, is the largest tributary to Lake Baikal and comprises ~82% of the lake's watershed [64] (Figure 1). The Selenga River supplies approximately 50% of water runoff and 60% of sediments to Lake Baikal [65]. Lake Baikal, the deepest and most voluminous freshwater lake in the world, holds approximately 20% of all liquid freshwater on Earth [63]. Located at a relatively high latitude in a semi-arid environment, the Selenga River Delta is particularly sensitive and vulnerable to climate change and water abstraction [66,67].

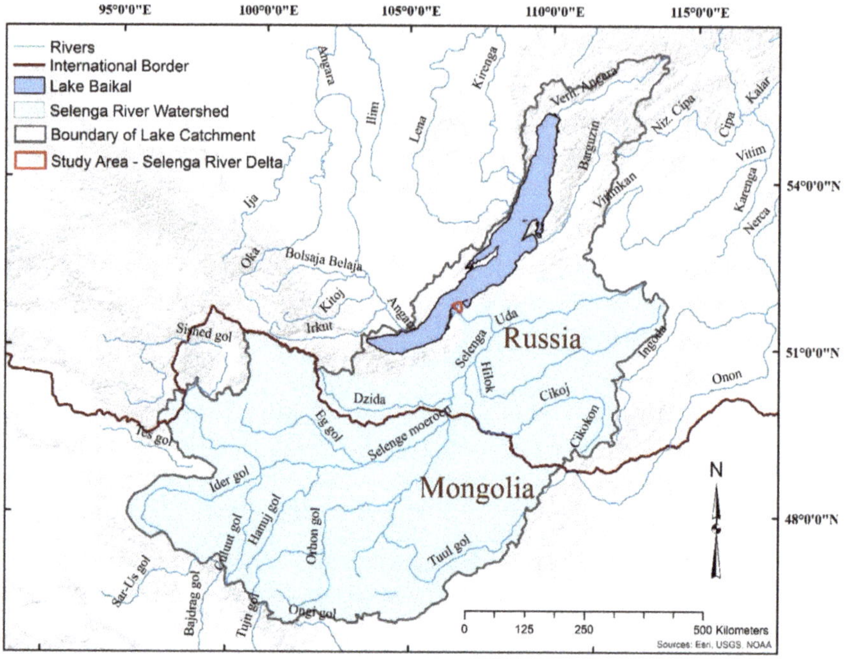

Figure 1. The watershed contributing to the study area, the Selenga River Delta into Lake Baikal, Russia.

2.2. Spatial Data, Preprocessing, and Initial Field Classifications

Two cloud-free Ortho-Ready Standard (OR2A) WorldView-2 (WV2) images from 25 June 2011, and 3 July 2011, were acquired for this study. There is a 5-km-wide overlap area between the 2 images. The WV2 multispectral imagery has 8 spectral bands (2.0 m pixel size): coastal (400–450 nm), blue (450–510 nm), green (510–580 nm), yellow (585–625 nm), red (630–690 nm), red-edge (705–745 nm), near infrared-1 (NIR1, 770–895 nm), and near infrared-2 (NIR2, 860–1040 nm), as well as a panchromatic band (0.5 m pixel size).

During the 2011 and 2012 field expeditions (described below), we collected 21 ground control points (GCPs) corresponding to relatively permanent features easily identified on the WV2 images, such as building corners, single isolated trees, "corners" (i.e., high curvature points) of river channels, and tree stands. GCP location data were collected using the Trimble Nomad and Yuma GPS receivers (Trimble Navigation Limited, Westminster, CO, USA) with 100 points averaged for each location. The locational geo-accuracy error of the WV2 images was less than 5 m as confirmed by the 21 GCPs. To make the 2 images radiometrically comparable, the digital number values of the original WV2 images were calibrated and converted to the top-of-the-atmosphere reflectance values, which accounts for solar geometry differences at the 2 image acquisition dates. The combination of the 2 images covers an area of 215 km^2, including the entire Kabansky Nature Reserve and its surrounding area (Figure 2). The WV2 data were not ortho-rectified.

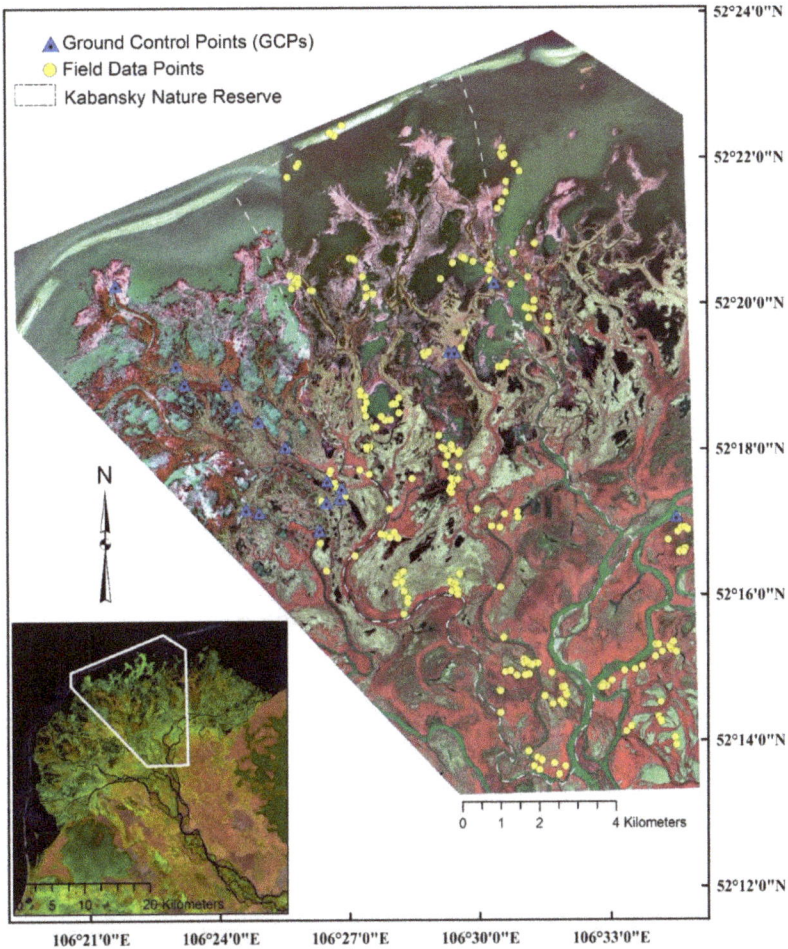

Figure 2. WorldView-2 false-color composite (near infrared-1 (NIR1), red, green) of the study area showing the spatial location of the field-sampling locations and ground control points. Inset image shows the Selenga River Delta in Lake Baikal and the study area boundary.

We conducted a preliminary ISODATA classification to inform our field study in 2011, initially selecting 24 classes with sufficiently high Jeffries-Matusita (J-M) separability values [68]. After the field expedition in 2011, we conducted a maximum likelihood classification to inform the 2012 field expedition, merging some classes to create a 22-class wetland landscape, as detailed in Lane et al. [60,61].

2.3. Field Data

A total of 228 field data points were collected over the 2011 and 2012 expeditions (see Figure 2). The field data were collected in July 2011 and July 2012. Homogeneous polygons derived from the preliminary classifications (described in Section 2.2) were visited by teams of botanists and ecologists. At least 3 unique polygons for each class were sampled across the study area, accessed by boat, hiking, and/or wading. Vegetation and structural data were collected at each point within a 100 m^2 sampling frame established for each sampling point. Information collected included identification of the species and relative abundance of all plants with $\geq$10% coverage and ancillary information, including water depth, nonvegetation abundance (i.e., bare ground), and substrate composition within the sampling frame. Photographs were taken from the center of each polygon in the 4 cardinal directions. No major changes occurred in the wetland systems between the 2 field data collection periods. For this analysis, species-level abundance data were collapsed to the genus level to facilitate comparisons among the classifiers.

2.4. Regions of Interest

Subsequent to the field expeditions, regions of interest (ROIs) were established using ENVI (Harris Geospatial Solutions, Herndon, VA, USA, version 5.3) around each field point (i.e., the center of each polygon) for use in the classification. Each ROI was approximately 28 pixels in size to approximate the field sampling frame, and was normally centered on the field point. Occasionally, an ROI was moved slightly off the field point to account for spectral noise and/or speckling. The pixels of each ROI were ascribed that ROI's value for the wetland class (i.e., one of 22 classes based on the 2012 field analyses) and assigned the spectral and ancillary data of the pixel values described in Section 2.5. We then randomly sampled ROIs for training (n = 158) and testing (n = 70) datasets. Each wetland class had, on average, 7 ROIs for training and 3 for testing (training range: 2–12 ROIs; testing range: 1–5 ROIs). The partition of training and testing datasets was performed by visually inspecting the point distribution so that the training and/or testing datasets were distributed in space to maximize the distance between the points, thereby minimizing the chances of spatial autocorrelation. A total of 6262 pixels for training and 2773 pixels for testing were identified.

2.5. Creating Spectral Metrics

Various spectral and landscape metrics can increase our ability to accurately discretize the landscape [69,70]. We therefore calculated additional characteristics to parametrize our models. Univariate Pearson's product-moment linear correlation analyses among the variables were also conducted (Table 1). Three spectral metrics expected to improve classification accuracy were calculated based on differing ratios between spectral bands: the Normalized Difference Vegetation Index (NDVI, [71]), Normalized Difference Water Index (NDWI, [72]), and Normalized Difference Soil Index (NDSI, [73]).

The NDVI is a well-established indicator for the presence and condition (i.e., abundance, vigor, and health) of vegetation [71]. The radiometrically calibrated reflectances of WV2 band 5 (Red) and band 7 (NIR1) were used to compute the NDVI. Healthy and abundant vegetation reflects strongly in the near-infrared portion of the spectrum while absorbing strongly in the visible red light portion, yielding high positive NDVI values. Sparse, stressed, and flooded vegetation has smaller positive NDVI values. Open water bodies yield negative values due to larger red reflectance than NIR. The NDVI values for bare soil ground are near zero due to their similar reflectance in both bands. The NDVI value in the study area ranges from −0.46 to 0.87 (data not shown).

We also calculated the NDWI following Wolf [73], based on the reflectance of WV2 band 1 (B1, Coastal) and band 8 (B8, NIR2) as $((B1 - B8)/(B1 + B8))$. Water features in the NDWI will typically have positive values, while soil and vegetation will have negative or zero values [74]. NDWI values in the study area range from −0.61 to 0.83 (data not shown).

Table 1. Correlation matrix of predictor variables. Linear correlations ($|r| \geq 0.89$) are shaded gray.

Predictor	Coastal (B1)	Blue (B2)	Green (B3)	Yellow (B4)	Red (B5)	Red-Edge (B6)	NIR1 (B7)	NIR2 (B8)	NDVI	NDSI	NDWI	LP	SD	DTS	Texture
Blue (B2)	0.96														
Green (B3)	0.87	0.88													
Yellow (B4)	0.93	0.96	0.93												
Red (B5)	0.87	0.95	0.79	0.94											
Red Edge (B6)	0.42	0.46	0.75	0.57	0.40										
NIR1 (B7)	0.21	0.27	0.56	0.38	0.26	0.95									
NIR2 (B8)	0.21	0.29	0.55	0.39	0.29	0.93	0.99								
NDVI	−0.06	−0.02	0.29	0.09	−0.03	0.79	0.86	0.86							
NDSI	−0.65	−0.69	−0.41	−0.70	−0.80	0.02	0.15	0.10	0.36						
NDWI	−0.21	−0.28	−0.52	−0.39	−0.30	−0.89	−0.94	−0.96	−0.92	−0.07					
LP	−0.21	−0.09	−0.07	−0.05	0.02	0.20	0.29	0.31	0.30	0.05	−0.30				
SD	0.24	0.12	0.11	0.09	0.00	−0.16	−0.25	−0.28	−0.37	−0.05	0.34	−0.50			
DTS	0.24	0.12	0.10	0.09	−0.01	−0.18	−0.28	−0.31	−0.39	−0.06	0.36	−0.50	0.99		
Texture	0.03	0.04	−0.18	−0.02	0.10	−0.48	−0.5	−0.48	−0.61	−0.31	0.53	−0.05	0.11	0.14	
DEM	−0.17	−0.05	−0.06	−0.03	0.06	0.16	0.27	0.31	0.28	0.00	−0.34	0.48	−0.54	−0.53	−0.15

Note: LP = landscape position; SD = surface depression; DTS = distance to stream; NDVI = Normalized Difference Vegetation Index; NDSI = Normalized Difference Soil Index; NDWI = Normalized Difference Water Index; DEM = Digital Elevation Model.

The NDSI discriminates soils from other background objects and was computed using WV2 shortwave bands 3 (B3, green) and 4 (B4, yellow; [73]) as ((B3 − B4)/(B3 + B4)). It is used to quantify dissimilarities of vegetation cover density and soil properties [75]. Despite the need for the shortwave infrared (SWIR) band to generate an index capable of discriminating soils from other objects (i.e., brightness intensity of soils is higher at longer wavelengths), the ratio of the aforementioned two visible bands of the WV2 dataset was consistently found to be a robust index in similar applications capable of discriminating soil from other image objects [73,76]. Exposed soils, non-photosynthesizing vegetation, and inundated areas tend to exhibit negative and low positive NDSI values, while healthy vegetated areas have higher positive NDSI values. For the study area, the NDSI ranges from −0.41 to 0.54 (data not shown).

We also calculated a texture metric [77] to characterize the spatial structure of the wetland vegetation and habitat [78]. We quantified the homogeneity texture variable using Gray Level Co-occurrence Matrices [79] with a grayscale quantification level of 64 and a 3 × 3 processing window. Water surfaces and emergent grasses have relatively smooth texture and subsequently high homogeneity values. Forested and scrub-shrub habitats have relatively rough texture and low homogeneity values, providing a contrast in habitat classification. Textural values in the study area range from 0.0 to 1.0 (data not shown).

2.6. Landscape Metrics and Topographic Data

To represent spatial associations and domain knowledge [69,70], we also introduced three hydrogeomorphic metrics: landscape/topographic position, distributary stream channels, and surface depressions. We further assigned elevation data to each ROI.

2.6.1. Landscape/Topographic Position Variable

The Selenga River Delta has a protruding fan shape with a loam- and sand-rich soil. The outer boundary of the delta is configured by a chain of long sand spits (depositional sandbars) (see Figure 2). Between the sandbar chain and the vegetated delta front is a subaqueous deltaic plain composed of silt and clay sediments. Hydrogeomorphologically, the delta is composed of low, central, and high portions. The low, northern, peripheral portion of the delta is subjected to regular floods [80]. The central floodplain and southern high islands are only flooded during high floods, and the floodplain terraces are typically not affected by floodwater. The floodplain terraces have flat or slightly undulating surfaces, which is complicated by the natural levees along former and present river channels [80].

We used the depositional sandbars described above to delineate the peripheral "shoreline" extent for the Selenga River Delta. Then the shortest distance from each WV2 pixel to the delta shoreline was created as an ancillary geographic information system (GIS) layer to indicate the landscape/topographic position along the relevant gradient in the broader landscape from lowlands, to midlands or central areas, to highlands. This variable is also closely related to the magnitude of seiches created from strong prevailing winds across the lake.

2.6.2. Distance to Stream Channels

The Selenga River enters Lake Baikal via the Selenga Delta wetland through a complex array of channels that are active and ice-free for approximately six months of the year. The inter-distributary bays are filled with sandy deposits frequently redistributed by waves. The delta consists of sandy lobe islands separated by numerous elongated and bifurcated channels [81]. Numerous distributary channels with natural levees separate extensive marshes, and lakes, channel cutoffs, and oxbows are abundant on the lobate islands. Natural levees with loamy sandy and sandy alluvial deposits and meadow marshes are widespread within the lower portion of the delta.

We extracted the stream channel features from the multispectral WV2 imagery sharpened by its panchromatic band using the NDWI thresholding method. The shortest distance from each WV2 pixel to the stream channels was created as another ancillary GIS data layer. The distance to distributary channels was used as a hydrogeomorphic variable to indicate the spatial relationship (proximity) to stream levees, closely related to elevation and flooding frequency, which has a controlling effect on wetland vegetation.

2.6.3. Distance to Depressional Features

Ponds, lakes, and marshes were extracted from the pan-sharpened WV2 imagery to indicate proximity to hydrogeomorphic features characterized as surface depressions. Depressions (or the ponds, lakes, and marshes that form in areas relatively distal to the stream network) are formed through ice-block and water-scouring actions, the magnitude of which is controlled by the energy of the water. We hypothesized that distance to a surface depression would be a hydrogeomorphic proxy for a combination of exposure to water-energy and sediment transport and could inform wetland vegetation typology. Similar to the stream channel extraction, we extracted the depressional features (e.g., lakes, ponds) using the NDWI thresholding method along with shape indices (see, e.g., [82]). The shortest distance from each WV2 pixel to the depressional features was created as another ancillary GIS data layer.

2.6.4. Surface Elevation

A digital surface-elevation dataset from the Advanced Spaceborne Thermal Emission and Reflection Radiometer-Global Digital Elevation Model (ASTER-GDEM, 30 m spatial resolution) was used as an auxiliary predictor variable. As with the aforementioned landscape metrics (predictors), elevation was expected to be a proxy for wetness, with areas of lower elevation being prone to longer hydroperiods. The dataset was projected (WGS84, UTM Zone 48 Northern Hemisphere) and resampled to a WV2 native pixel resolution of 2 m. The elevation of the study area ranged from 419 to 484 m above sea level, with an average elevation of 446 m.

2.7. Decision-Tree, Rule-Based, and Random Forest Classification and Assessment

2.7.1. Overview

We contrasted three different classification approaches: DT, RB, and RF. The classification models were constructed iteratively, adding information content at each step. Each iteration was considered a "test," and the overall accuracy of that iteration and approach contrasted between the three methods. Models were initially constructed using the four traditional bands: Test 1: B2 (Blue), B3 (Green), B5 (Red), B7 (NIR1). We then added the WV2 coastal band (B1) as Test 2, as Lane et al. [60] determined that this band could facilitate open-water and vegetated habitat discrimination in wetlands. The full eight bands available in WV2 were analyzed (Test 3), and we then iteratively augmented the eight-band stack in a stepwise approach with the derived indices:B1–8 plus Test 4: NDVI, Test 5: NDWI, Test 6: NDSI, Test 7: texture, and Test 8: elevation. We tested WV2 B1–B8 plus all spectral metrics (e.g., NDVI, NDWI, etc.) initially without boosting (Test 9), and then adding the "boost" function described below (Test 10) for the DT and RB approaches only. We analyzed B1–B8 plus the four spectral metrics and all three hydrogeomorphic variables (Test 11); we subsequently analyzed the same 15-layer stack using the "boost" function for the DT and RB approaches (Test 12). We analyzed B1–B8 plus the four spectral metrics and all three hydrogeomorphic variables (Test 11). We subsequently analyzed the same 15-layer stack using the "boost" function for the DT and RB approaches (Test 12). We analyzed B1–B8 plus four spectral metrics, three hydrogeomorphic variables, and elevation initially without boosting (Test 13), and with boosting for the DT and RB approaches only (Test 14). Lastly, since we intended to develop classification models with the minimum number of input variables possible (thereby decreasing data dimensionality) while achieving the highest possible overall accuracy, we systematically worked through model parameterization to develop the most parsimonious models by removing highly correlated variables (see Table 1; Tests 15 and 16).

2.7.2. Decision-Tree Classification

Both DT and RB models were developed using C5.0 [83]. DT classification employs a hierarchy of rules in an automated, top-down, dichotomous fashion [84–86]. C5.0 uses a recursive partition procedure to build a binary DT. The DT is composed of different levels of nodes: a root node, a set of internal nodes (branches), and a set of terminal nodes (leaves). The root node or whole dataset is divided (split) into more homogeneous groups. The split at each internal node of a tree is defined by a single predictor variable based on statistical analysis of the training data [43]. The split is made based on determining predictor variable values with the most discriminatory power, measured by the information gain ratio with iterative predictor-variable values in C5.0 [36]. Split nodes subsequently contain only part of the data and can further be divided until an end node (leaf) is reached, where no further split is possible or desired. Both DT and RB models can be improved through boosting, wherein misclassified leaves in the final model are reanalyzed and the model reiteratively runs in an attempt to properly classify these errors.

The DT was constructed in two steps. First, a large tree was grown to fit the training data closely. Then, the tree was pruned (or winnowed) to remove attributes that affected the error rate. This pruning process attempted to correct overfitting errors and reduce the tree size [36,83,87]. DT and RB models were boosted in Tests 10, 12, 14, and 16, wherein the misclassified leaves in the final DT were reanalyzed and the model was reiteratively run in an attempt to properly classify these errors. Boosting was conducted for 10 trials or until the model performance failed to improve as measured by the error rate. The portion of the DT from the 5-layer stack (WV2 coastal, green, red, and NIR1 bands plus texture) is displayed as a branching dichotomous tree in Figure 3.

```
NIR1 <= 165:
:...NIR1 <= 55:
:   :...Red > 60:
:   :   :...Red <=73: 3 (189/1)
:   :   :   Red > 73:
:   :   :   :...Green > 98: 3 (108/1)
:   :   :       Green <= 98:
:   :   :       :...Coastal <= 124: 3 (5)
:   :   :           Coastal > 124: 4 (52)
:   :   Red <= 60:
:   :   :...Red <=45:
:   :       :...NIR1 <= 47:
:   :       :   :...NIR1 <= 32:
:   :       :   :   :...Green <= 69: 6 (5)
:   :       :   :   :   Green  > 69: 7 (6)
:   :       :   :   NIR1 > 32:
:   :       :   :   :...Coastal > 119: 6 (236)
:   :       :   :       Coastal <= 119:
:   :       :   :       :...Texture <= 0.94444: 6 (35/2)
:   :       :   :           Texture > 0.94444:
:   :       :   :           :...Red <= 43: 6 (2)
:   :       :   :               Red > 43: 2 (4)
:   :       :   NIR1 > 47:
:   :       :   :...Texture <= 0.6265: 9 (8/1)
:   :       :       Texture > 0.6265:
:   :       :       :...Red <= 42: 8 (14)
:   :       :           Red > 42:
:   :       :           :...NIR1 > 53: 8 (3/1)
:   :       :               NIR1 <= 53:
:   :       :               :...Coastal <=122: 6 (14)
:   :       :                   Coastal > 122:
:   :       :                   :...NIR1 > 49: 6 (4/1)
:   :       :                       NIR1 > 49: 8 (3)
```

Figure 3. An example of the decision-tree outcome for classifying wetlands of the study area.

2.7.3. Rule-Based Classification

The DT developed in 2.7.2 (including the boosting conducted in Tests 10, 12, 14, and 16) was transformed into a simpler set of "if-then" rules in C5.0 by creating "rulesets" in the algorithm. The ruleset generated from the DT has fewer rules than the number of leaves in the decision tree, thus it is a more compact and simpler representation [88]. Since each conditional logic rule describes a specific context associated with a class, it is relatively easy to examine, validate, and interpret the ruleset. The "if-then" logic rules make the connection between wetland classes and their predictor variables. A portion of the ruleset when using the same five-layer stack as in Figure 3 is shown in Figure 4.

```
Rule 33: (119/1, lift 18.4)      Rule 37: (2, lift 14.1)
          NIR1 > 62                        Texture > 0.33098
          NIR1 <= 117                      Texture <= 0.3732
          Red <= 43                        NIR1 > 165
          Green <= 73                      NIR1 <= 221
          Coastal <= 122                   Red > 52
          → class 8[0.983]                 Coastal <= 122
                                           → class 8[0.750]
Rule 34: (66/7, lift 16.5)       Rule 38: (9/5, lift 8.5)
          NIR1 > 77                        Texture > 0.33098
          NIR1 <= 122                      Texture <= 0.34209
          Red <= 48                        NIR1 <= 242
          Coastal <= 121                   Coastal > 122
          → class 8[0.882]                 → class 8[0.455]
Rule 35: (4, lift 15.6)          Rule 39: (1450/1129, lift 4.2)
          Texture > 0.37163                NIR1 <= 177
          Texture <= 0.49542               Red <= 53
          NIR1 > 122                       → class 8[0.222]
          NIR1 <= 165            Rule 40: (15, lift 30.5)
          Red <= 56                        Texture <= 0.5732
          Coastal > 120                    NIR1 > 60
          Costal <= 122                    NIR1 <= 102
          → class 8[0.933]                 Red > 43
Rule 36: (3, lift 15.0)                    Red <= 51
          Texture <= 0.34993               Green <= 73
          NIR1 <= 165                      Coastal > 120
          Red <= 58                        → class 9[0.941]
          Green > 88
          Coastal <= 124
          → class 8[0.800]
```

Figure 4. An example of the rule-based approach for classifying the Selenga River Delta wetlands.

2.7.4. Random Forest Classification

In contrast with the single optimal tree built using the entire training dataset and all of the predictor variables in the DT and RB approaches, RF creates an ensemble of trees that each provides a "vote" to select the best classification approach. That is, class membership in a DT is decided by a single tree, whereas the majority of votes from the assemblages of trees built by RF decide the class assignment of a given pixel. We used the randomForest package [51] in the R statistical software environment (RStudio, Inc., Boston, MA, USA, version 1.0.143). Each RF tree was built by training each DT (ntree) with a random subset of the predictor-variables (mtry) from the training dataset with a replacement [49,89]. Based on the preliminary analyses, we selected RF models comprising ntree = 1000 trees, 1000 bootstrap (or "out-of-box", OOB) samples to assess internal model error, and tested multiple predictor variables (i.e., mtry = the square root of the total number of input variables, either 2, 3, or 4) at each split as we tested from the simplest to the most complex model (i.e., 4-, 9-, and 16-layer stacks).

2.7.5. Accuracy Assessment

Each ROI pixel in the training dataset (i.e., 158 ROIs composed of 6262 pixels) was used to construct DT, RB, and RF models, ranging from the relatively simple (i.e., four bands as input) to complex (i.e., 16 data layers including eight bands, four spectral metrics, three hydrogeomorphic metrics, and elevation). The holdout 70 ROI (2773 pixels) as a validation dataset was used to assess the prediction accuracy of the approach across all 16 tests, resulting in an average of 103 validation pixels generated per class for each of the 22 classes in the evaluation. Performance measures included overall accuracy, class-wise producer's accuracy (PA; errors of omission), and class-wise user's accuracy (UA; errors of commission). We quantitatively assessed if the observed difference in the classification accuracies between the "best" or most accurate application of each of the approaches was statistically meaningful using 95% confidence intervals [90].

3. Results

3.1. Field Data Collection

Fifty-two different plant genera were found at ≥10% coverage in our 228 sampling frames. Members of the genera *Equisetum* and *Carex* were most commonly found, with 58 and 45 sites, respectively. *Nymphoides* (41 sites) and *Salix* (27 sites) were also commonly encountered. Open water (100 sites) and thatch (45 sites) were also noted to cover ≥10% of the sampling frame in the field sites. Fourteen genera were encountered only a single time. Because the main goal of this paper is to determine the effectiveness of the DT, RB, and RF approaches, we do not further describe the ecology of the wetland classes here (but see [60,61]).

3.2. Decision-Tree, Rule-Based, and Random Forest Classification Accuracy and Complexity

We examined the performance of the DT, RB, and RF approaches based on a random sample of 2773 validation pixels, determined from field-sampled sites and independent of the training pixels. Performance measures included OA, class-wise PA (errors of omission), and class-wise UA (errors of commission).

3.2.1. Classification Accuracy

The training and testing dataset ROIs were independent and widely distributed across the study area, minimizing the potential for spatial autocorrelation by predictor variables. As shown in Table 2, the OA on the testing dataset ranged from 54.8 (Test 11: 15-layer stack and RB classification) to 81.2% (Test 16: 5-layer stack and RF classification). The highest OAs for DT and RB were 80.7% (Test 16) and 80.0% (Test 10), respectively. Both tests achieved the highest accuracy with boosted classification. Test 16 was also the highest-performing RF classification (81.2% OA). We assessed the classification accuracy of the best-performing models for each approach, and, as the 95% confidence intervals overlapped, we found no significant differences between the test results (DT Test 16, RB Test 10, and RF Test 16).

Table 2. Contrasting the accuracy between three classifiers and combinations of predicator variables. The highest overall accuracy (OA) per test is identified with bold font.

Test	Input Layers	Decision-Tree (DT) Classification				Rule-Based (RB) Classification				Random Forest (RF) Classification		
		Training Data		OA on Testing Data		Training Data		OA on Testing Data		Training Data	OA on Testing Data	
		# Tree Leaves	Error (%)	Mean (%)	95% CI	# "If-Then" Rules	Error (%)	Mean (%)	95% CI	Out-of-Box Error (%)	Mean (%)	95% CI
1	4 traditional bands (B2 + B3 + B7)	222	6.7	66.9	65.1–68.7	136	7.3	66.5	64.7–68.3	9.9	**73.1**	71.4–74.7
2	5 traditional bands (B1 + B2 + B3 + B5 + B7)	252	5.4	69.2	67.4–70.1	157	6.1	66.6	64.8–68.4	8.8	**74.0**	72.4–75.7
3	8 traditional bands (B1-B8)	270	3.5	73.1	71.4–74.7	168	4.0	74.7	73.0–76.3	6.6	**76.7**	75.1–78.2
4	8 traditional bands + NDVI	255	3.4	73.4	71.7–75.0	161	4.0	73.5	71.8–75.1	6.8	**75.7**	74.0–77.3
5	8 traditional bands + NDWI	251	3.5	72.8	71.1–74.5	152	4.1	73.6	72.0–75.3	6.6	**77.0**	75.4–78.6
6	8 traditional bands + NDSI	219	3.5	73.1	71.4–74.7	165	3.8	71.9	70.2–73.6	6.6	**77.0**	75.4–78.6
7	8 traditional bands + texture	226	2.7	78.2	76.6–79.7	156	3.1	77.7	76.1–79.2	4.9	**81.1**	79.6–82.6
8	8 traditional bands + elevation dataset	139	2.2	61.6	59.8–63.4	162	2.5	61.9	60.0–63.7	4.2	**75.3**	73.6–76.9
9	8 traditional bands + 4 indices (NDVI, NDWI, NDSI, texture)	225	2.4	77.2	75.6–78.8	140	2.9	78.7	77.2–80.2	5.1	**80.6**	79.0–82.0
10	8 traditional bands + 4 spectral indices; with boost (10 trials)	Boost	0.1	80.1	78.6–81.6	Boost	0.0	80.0	78.5–81.5			
11	8 traditional bands + 4 spectral indices + 3 hydrogeomorphology variables	49	0.8	55.3	53.4–57.1	48	0.8	54.8	52.9–56.6	1.6	**74.7**	73.1–76.3
12	8 traditional bands + 4 spectral indices + 3 hydrogeomorphology variables; with boost (10 trials)	Boost	0.0	60.2	58.3–62.0	Boost	0.0	58.9	57.1–60.8			
13	8 traditional bands + 4 spectral indices + 3 hydrogeomorphology variables + elevation dataset	153	0.7	58.0	56.1–59.8	100	0.8	58.3	56.4–60.1	1.3	**73.0**	71.2–74.5
14	8 traditional bands + 4 spectral indices + 3 hydro attributes + elevation dataset; with boost (10 trials)	Boost	0.0	63.1	61.3–64.9	Boost	0.0	59.9	58.0–61.7			
15	Uncorrelated and parsimonious (B1 + B3 + B5 + B7 + texture)	221	3.4	75.7	74.0–77.3	154	3.9	74.0	72.4–75.7	6.2	**81.2**	79.7–82.6
16	Uncorrelated and parsimonious (B1 + B3 + B5 + B7 + texture) with boost	Boost	0.7	80.7	79.2–82.1	Boost	0.7	77.8	76.2–79.3			

13

3.2.2. The Effects of Additional Bands and Input Parameters

Similar to Lane et al. [60], we found that the addition of available spectral bands in WV2 increased OA (Tests 1 and 3) across the DT, RB, and RF models by 6.2%, 8.2%, and 3.6%, respectively. The improvements from adding derived indices (i.e., NDVI, NDWI, NDSI) vacillated across the three approaches, with no marked increase in overall accuracy (e.g., Tests 4 to 6), but adding texture increased OA by 3–5%, depending on the approach (contrasting Test 3 with Test 7). As the 22 wetland classes were composed of different hydroperiods and inundation regimes, vegetative structures, and soil characteristics that affect the spectral signal received by the WV2 sensors, particular classes (e.g., those with abundant forest structure) might respond with substantial increases (or decreases) in comparative accuracy between different tests.

However, contrary to our expectations, including hydrogeomorphic variables and/or elevation data resulted in a marked decrease in OA across all approaches (e.g., contrast Test 3, Test 8, and Test 11 with Test 14 across all approaches in Table 2). For example, including the elevation dataset decreased the OA by 11.5%, 12.8%, and 1.4% for DT, RB, and RF, respectively (e.g., contrast Test 3 and Test 8 in Table 2). Similarly, including three hydrogeomorphic variables (landscape position, distance to stream channels, and distance to depressional features; Sections 2.6.1–2.6.3) along with the eight multispectral bands and four spectral indices resulted in a decrease in OA by 21.9% for DT and 23.9% for RB (Test 9 and Test 11, Table 2). With the boost function, there were similar decreases in OA, 19.9% for DT and 21.1% for RB (Test 10 and Test 12). A similar decrease was also observed for RF, though by a smaller amount, 5.9% (see Table 2).

Removing highly correlated ($|r| \geq 0.89$) predictor variables (e.g., B2, blue; B4, yellow; B6, red-edge; and B8, NIR2; see Table 1) from Test 7 yielded approximately the same OA results as shown in Test 16 (see Table 2): 80.7% for DT and 81.2% for RF. Consequently, for DT and RF, the models that combined parsimony and accuracy were built using a five-layer stack of input variables (Test 16: B1, coastal; B3, green; B5, red; and B7, NIR1; and texture). Using the same parsimonious predictor variables, a lower OA of 77.8% was achieved for RB (i.e., in contrast with Test 10, 80.0% RB OA).

4. Discussion

4.1. Random Forest as the Classifier of Choice

In all iterations (see Table 2), the RF model outperformed both DT and RB. In addition, the RF model appeared to better handle an increasing number of predictor variables that resulted in higher OA, demonstrating its ability to effectively process complex and highly dimensional datasets. Furthermore, RF provides useful information to the end user in terms of mean decrease in Gini (MDG, [49]), a measure of the relative importance of different predictor variables affecting overall accuracy. Though not the focus of this study, the MDG indicates that NIR1 (B7) has the greatest effect on overall model accuracy in the best RF model (Test 16, data not shown). Therefore, end users wishing to increase OA could consider ensuring that NIR1 (B7) is used in models addressing their study area, focusing the development of additional indices to improve OA on other information content in the spectral data (e.g., focusing on the effects of soil reflectance, or calculating texture metrics).

DT and RB models were able to approximate the RF results (e.g., OA between all three approaches was within 0.5% in Test 10 and 3.5% in Test 16), and overlapping confidence intervals indicated no significant differences between the different approaches [90]. DT and RB were only able to achieve near parity with RF through the use of the boost function in C5.0. Similar to the voting aspect of RF, the boost function predicts a given class assignment by using the majority of votes from multiple classifiers as opposed to a single tree or ruleset. Moreover, similar to the random selection of input variables (mtry) by RF, the ability to make subsets of the predictor variables to construct the DTs and rulesets was achieved through the "winnowing" mechanism of the C5.0 package. However, relative to RF, these different steps in the DT and RB approaches require additional processing and user input.

Throughout the literature we found many instances of users classifying landscapes with DT and RB methods, but we found only one, Rodriguez-Galiano et al. [91], that contrasted the outcome between a DT and RF. In their study, they mapped 14 land-cover categories using Landsat TM and ancillary datasets with OA of 86% for the DT approach. Similar to our efforts, RF increased OA to 92%, and RF also outperformed DT/RB when model parsimony was optimized.

However, RF is considered a "black box" model, wherein much of the algorithm is performed in the virtual background [92]. That requires accepting the outcomes or laborious efforts to unpack the algorithm. In addition, RF OA can be sensitive to the distribution of ROIs; unequal distribution can affect the RF outcome relative to a balanced approach [93,94]. However, it appears the benefits of RF outweigh the detriments, and as such we suggest that end users strongly consider using RF in their classification applications.

DT and RB may be useful when the rules and/or tree nodes and splits are contextually relevant and useful to end users [84], versus the aforementioned "black box" nature of RF resulting in difficult rule extraction and model comprehension. Should end users rely on DT and RB approaches, we recommend using the boost and winnowing functions of the C5.0 package, which could improve the OA of the DT and RB classifications.

4.2. Overall Accuracy with a Large Suite of Classes

Whereas land cover classification is common, using remotely sensed data to specifically assess and conduct wetland classification is somewhat rarer (see, e.g., [95] for a detailed review of approaches). Furthermore, we have found that most wetland classifications limit the classes to a relatively small number, depending on the end goals of the user (e.g., 10 to 11 classes, [9,96]). The deltaic wetland we studied was discriminated into 22 classes, which set a high bar for achieving acceptable overall classification accuracy. We would have expected higher OA if we had fewer classes, or if we targeted certain classes and perhaps merged them together.

The 22 classes had high Jeffries-Matusita (J-M) separability values [68] Table 3). These values range from 0.00 to 2.00, with values <1.0 suggesting poor separability and values approaching 2.0 indicating high separability; we found 18 contrasts with J-M values less than an arbitrary 1.75, and only two instances where the J-M values were <1.20 (Class 8 and Class 9; Class 21 and Class 22). As evidenced by the above findings, lower J-M values typically occurred along neighboring classes, as might be expected. Dubeauet et al. [97] classified a headwater wetland ecosystem in the Dabus River basin, a large tributary of the Abay-Blue-Nile River in Ethiopia, using Landsat TM and an RF approach. Similar to our findings, they found that among the eight wetland types and three upland classes, the greatest confusion was between similar neighboring plant types and vegetation structures (e.g., greater confusion within herbaceous classes than between herbaceous and woody/shrub or open water classes). In general, our table-wide J-M average was 1.95, suggesting that the 22 classes were well discriminated by the high-resolution WV2 data.

Table 3. Jeffries-Matusita (J-M) distance measure of class separability for input predictor variables combination of WV2 bands 1–8; the table-wide J-M value was 1.95, indicating high overall separability.

Wetland Class	1	2	3	4	5	6	7	8	9	10	11	12	13	14	15	16	17	18	19	20	21
2	1.95																				
3	2.00	1.96																			
4	2.00	1.99	1.69																		
5	2.00	2.00	2.00	2.00																	
6	2.00	1.86	2.00	2.00	2.00																
7	1.94	1.94	2.00	2.00	2.00	1.98															
8	2.00	1.80	2.00	2.00	2.00	1.54	1.99														
9	2.00	1.66	2.00	2.00	2.00	1.74	2.00	1.12													
10	1.99	1.83	2.00	2.00	2.00	1.86	1.95	1.91	1.89												
11	2.00	1.96	2.00	2.00	2.00	2.00	2.00	1.99	1.97	1.34											
12	2.00	1.99	2.00	2.00	2.00	2.00	2.00	2.00	1.98	1.90	1.64										
13	2.00	1.98	2.00	2.00	2.00	2.00	2.00	1.98	1.92	1.99	1.95	1.73									
14	2.00	1.88	2.00	2.00	2.00	2.00	2.00	1.87	1.52	2.00	2.00	2.00	1.98								
15	2.00	1.95	2.00	2.00	2.00	2.00	2.00	1.99	1.87	2.00	2.00	2.00	2.00	1.57							
16	2.00	1.97	2.00	2.00	2.00	2.00	2.00	2.00	1.89	2.00	2.00	2.00	2.00	1.82	1.70						
17	2.00	1.99	2.00	2.00	2.00	2.00	2.00	2.00	1.98	2.00	2.00	2.00	2.00	1.93	1.98	1.94					
18	2.00	1.98	2.00	2.00	2.00	2.00	2.00	2.00	1.97	2.00	2.00	2.00	2.00	1.98	2.00	1.94	1.70				
19	2.00	2.00	2.00	2.00	2.00	2.00	2.00	2.00	2.00	2.00	2.00	2.00	2.00	1.98	2.00	1.99	1.52	1.69			
20	2.00	1.98	2.00	2.00	2.00	2.00	2.00	2.00	1.93	2.00	2.00	2.00	1.98	1.98	2.00	1.95	1.98	1.79	1.99		
21	2.00	2.00	2.00	2.00	2.00	2.00	2.00	2.00	2.00	2.00	2.00	2.00	2.00	2.00	2.00	2.00	2.00	2.00	2.00	1.68	
22	2.00	1.96	2.00	2.00	2.00	2.00	2.00	1.94	1.74	2.00	2.00	2.00	1.95	1.88	1.95	1.84	1.99	1.95	2.00	1.05	1.71

4.3. Metrics, Classes, Spectral Bands, and Hydrogeomorphic Variables

Using vegetative, soil, and hydrologic indicators based on various band combinations (NDVI, NDSI, and NDWI, respectively) did not markedly improve the classification. This is similar to the findings of Berhane et al. [47], who reported decreased classification OA (and/or no meaningful change) when exploring the influence of over 30 predictor variables on classification accuracy (using Quickbird imagery), including the NDWI, NDVI, and a functionally similar soil metric, the NDSI. However, they did find increased OA when incorporating a metric functionally similar to the NDWI, the Water Ratio Index [98]. Thus, though the primary goal in this study was to contrast the three classification approaches, in order to fully and accurately characterize the wetland landscape, full consideration of a multitude of band combinations should be explored (see, e.g., [47], Table 1, for a list of potential metrics to consider; see also [95]).

The results of this study, as well as those of Franklin and Peddle [78], Lane et al. [60,61], Berhane et al. [47], and others, show that including the homogeneity texture variable, calculated using the Gray Level Co-occurrence Matrices [79], did substantially improve classification OA. Water surfaces and emergent grasses have relatively smooth texture and subsequently high homogeneity values. Increasing vegetative structure, such as that found in forested and shrub-scrub wetland habitats, imparts relatively rough texture and low homogeneity values, apparently providing a useful contrast in wetland vegetation and habitat classification. These structural features of different wetland vegetation types, as determined by the texture measure included in this study, are thus recommended to provide a useful and additive metric to improve classification outcomes.

There appears to be relatively strong discriminatory power among the spectral bands when used in combination; exploring the median distribution of the WV2 bands across the 22 classes (Figure 5) supports this statement. In other words, the fact that the derived metrics, notwithstanding texture, did not markedly improve OA appears to be a manageable situation, wherein these data alone can provide useful information to the end user. This is further supported by the relatively high ~75% OA for the eight WV2 bands when analyzed as an 8-band stack (Test 3, Table 2). For instance, NIR1 and NIR2 have values closer to zero for Classes 1–7 (excluding Class 4) and increase linearly through to Class 22, with relatively minor overlap between median reflectance values among the wetland classes in this study. Interestingly, the remaining six bands indicate that the classes may be visually discriminated into approximately four clusters: Classes 1–5, Classes 6–13, Classes 14–16, and Classes 17–22; these may be further explored to better understand wetland classification and ecology. For instance, we found that our classification followed a wet-to-dry gradient, as evidenced by the vegetation data in Figure 6 and more closely explored in the wet-to-dry, north-to-south gradient evidenced in Figure 7A–H. Deeper waters and submerged vegetation manifested in the northern portion of the study area, and emergent vegetation and facultative upland genera were found in the southern portions.

With this distribution of wetland classes following a wet-to-dry gradient, we were surprised that our hydrogeomorphic and elevation data did not improve the classification. Indeed, including these variables decreased overall accuracy (e.g., Test 9 versus Test 11, Table 2; see also Test 3 and Test 8 for elevation effects). The elevation effect may be ascribed to the relatively flat nature of the delta, where the granularity of the topographic terrain provided no meaningful or discernable attribute information about the wetland classes to the classifiers, but rather imparted noise. We had expected elevation to play an important role, as even slight differences in elevation can dramatically affect inundation patterning, soil biogeochemistry, and vegetative structure. Perhaps if a higher-resolution DEM was found for the study area, we might find it a useful input variable.

Similarly, including three hydrogeomorphic variables (landscape position, distance to stream channels, and distance to depressional features; Sections 2.6.1–2.6.3) decreased overall accuracy by approximately 20% for DT and RB and 5% for RF (e.g., Tests 9 and 11, and Tests 10 and12 in Table 2). The complexity of the delta, wherein stream networks and channels are constantly migrating, affecting inundation patterning and water clarity and modifying hydrologic gradients, likely also meant that the granularity of our hydrogeomorphic variables was too coarse. We suspect that further refinement of hydrogeomorphic variables may improve their influence on OA (e.g., incorporating stream width, flow direction, and proportion of flow, as an energy surrogate, may correlate with the distribution of the wetland habitats).

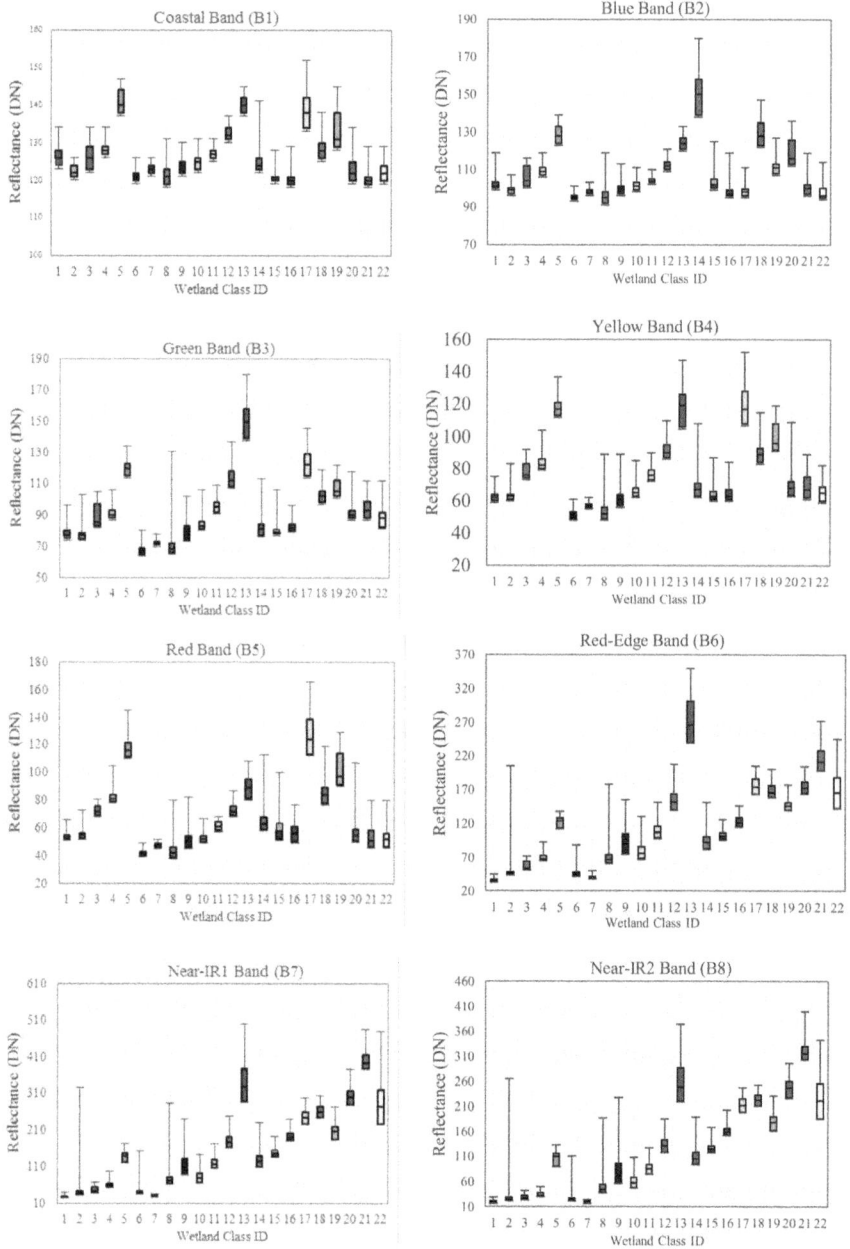

Figure 5. Median WV2 band distribution indicates strong discriminatory power between classes.

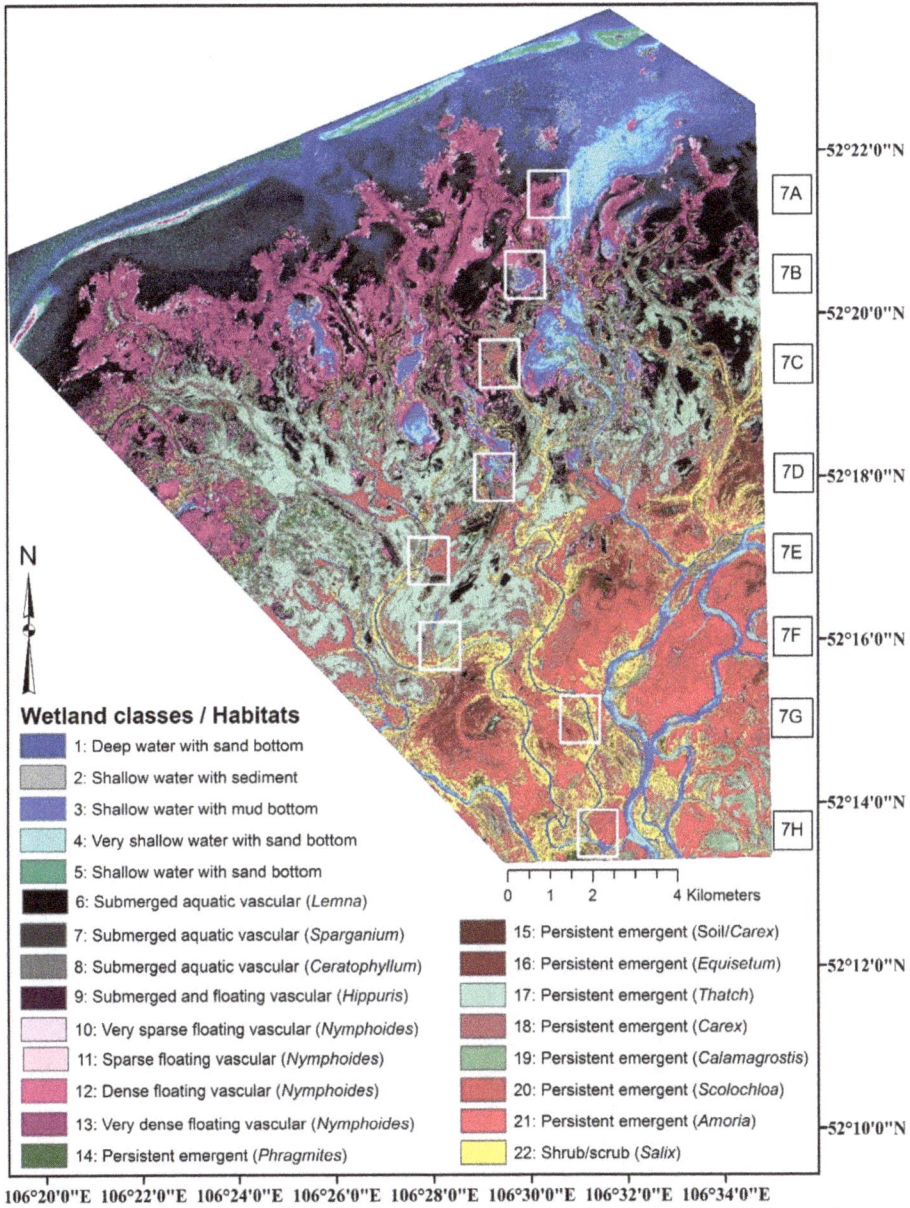

Figure 6. The RF-classified study area. Classes in the legend were attributed based on wetland plant abundance, water depth, and substrate composition (see, e.g., [60]). The north-to-south, wetter-to-drier boxes in Figure 6 are further discussed in Figure 7A–H.

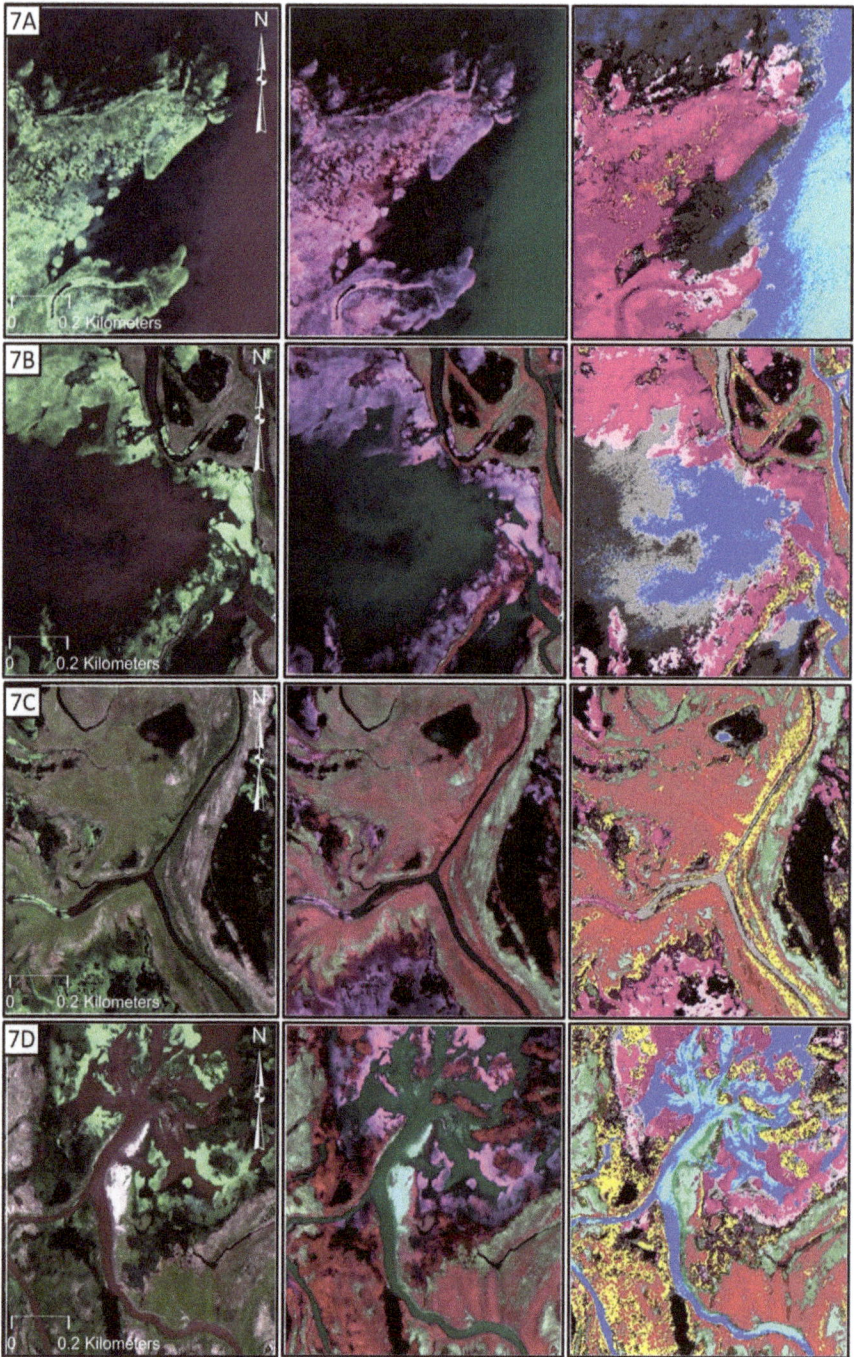

Figure 7. *Cont.*

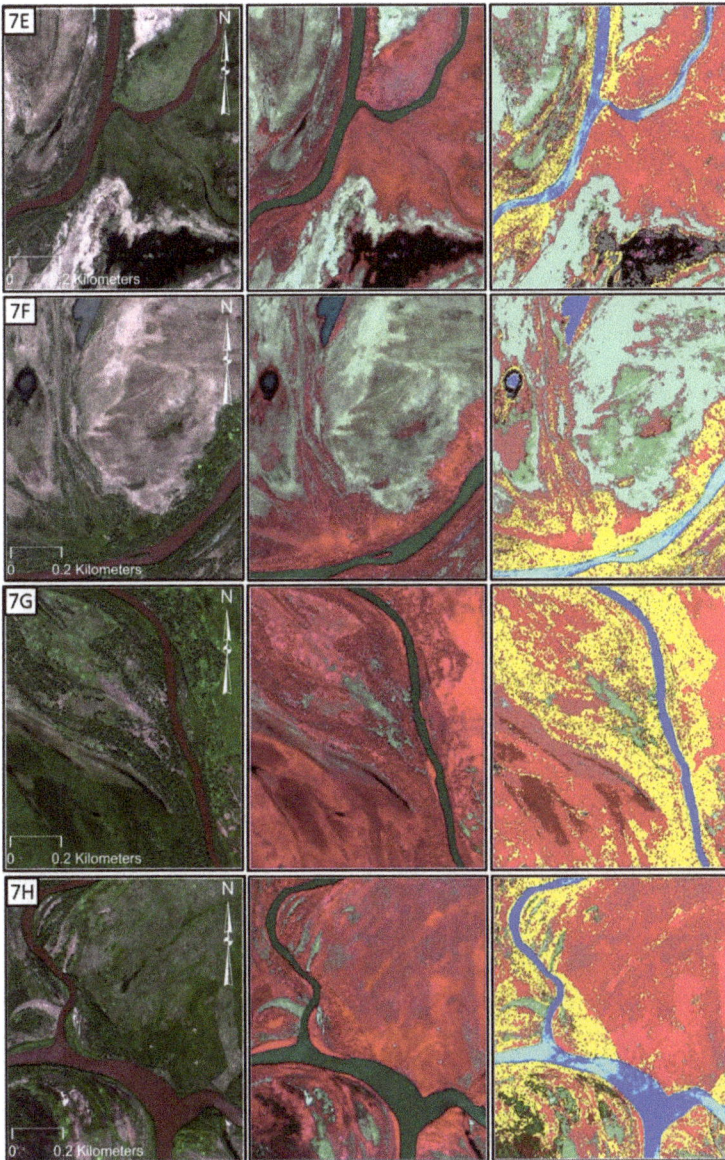

Figure 7. (**A–H**). The vegetation of the Selenga River Delta follows a north-to-south and wetter-to-drier gradient, as evidenced by the abundance of different wetland classes within the white rectangles in Figure 6. The images are combined WV2 bands 532 (**left**), bands 753 (**middle**), and the wetland classification thematic map (**right**) using the legend in Figure 6.

5. Conclusions

In this paper, we systematically and comprehensively evaluated the utility of three nonparametric machine-learning algorithms (DT, RB, and RF) for effective supervised classification of 22 complex freshwater deltaic wetland vegetation and aquatic habitats in the Selenga River Delta of Lake Baikal,

Russia. The use of WV2 multispectral bands, derived spectral indices, and ancillary data was optimized through iterative modeling and predictor variable selection to achieve a satisfyingly accurate working model. Our analysis shows that DT, RB, and RF classification methods provide a suitable framework to combine different types of data sources, accommodating image-derived indices and ancillary hydrogeomorphic variables in addition to image spectral bands and elevation datasets. The OA of the DT, RB, and RF classification methods ranged from 54.8 to 81.2%. The RF classification outperformed both the DT and RB classifications; performance is approximately equal when boost and winnowing functions, available in the C5.0 package, were used. We conclude that RF can be used as the classifier of choice in most cases, except, potentially, in situations where end users require narrative rules to best manage their resources. That would call for the DT or RB approach, though the breadth and abundance of rules (upwards of 140 rules or tree leaves to achieve OA $\geq$80%; see Table 2) may be daunting. Including a texture metric (homogeneity) substantially improved the classification OA. However, we were surprised that including vegetation/soil/water metrics (based on band combinations), hydrogeomorphic, and elevation data layers did not markedly improve OA. This may be a result of the complexity of the deltaic wetland system, which requires finer-resolution spatial data to meaningfully improve classification models.

Acknowledgments: This paper has been reviewed in accordance with the U.S. Environmental Protection Agency's peer-review policies and approved for publication. Mention of trade names or commercial products does not constitute endorsement of recommendation for use. Statements in this publication reflect the authors' personal views and opinions and should not be construed to represent any determination of policy of the U.S. Environmental Protection Agency. The research of O.A. Anenkhonov and V.V. Chepinoga was carried out using the framework of projects 0337-2016-0001 and 0347-2016-003 accordingly, supported by the Russian Federal Budget.

Author Contributions: All authors made significant contributions to the manuscript. Charles R. Lane developed and supervised the study, collected field data, analyzed the results, and co-wrote the manuscript. Tedros M. Berhane processed the spatial data, developed remote sensing methodologies, analyzed results, and co-wrote the manuscript. Qiusheng Wu provided methodological development, analyses, results interpretation, and manuscript additions and revisions. Oleg A. Anenkhonov and Victor V. Chepinoga co-developed the study's field component, provided botanical expertise in collecting and interpreting field data, and contributed to results interpretation and manuscript writing and revision. Brad C. Autrey co-developed the study, collected, processed, and analyzed field data, and contributed to results interpretation and manuscript editing and revision. Hongxing Liu conducted the initial spatial analyses and contributed to methodological development and results interpretation, as well as manuscript editing and revision.

Conflicts of Interest: The authors declare no conflict of interest.

References

1. Titus, J.; Hudgens, D.; Trescott, D.; Craghan, M.; Nuckols, W.; Hershner, C.; Kassakian, J.; Linn, C.; Merritt, P.; McCue, T. State and local governments plan for development of most land vulnerable to rising sea level along the US Atlantic coast. *Environ. Res. Lett.* **2009**, *4*, 044008. [CrossRef]
2. Klemas, V. Remote sensing of wetlands: Case studies comparing practical techniques. *J. Coast. Res.* **2011**, *27*, 418–427. [CrossRef]
3. Biggs, J.; von Fumetti, S.; Kelly-Quinn, M. The importance of small waterbodies for biodiversity and ecosystem services: Implications for policy makers. *Hydrobiologia* **2017**, *793*, 3–39. [CrossRef]
4. Mitsch, W.; Gosselink, J. *Wetlands*, 2nd ed.; Van Nostrand Reinhold Press: New York, NY, USA, 1993.
5. Finlayson, C.; Davidson, N.; Spiers, A.; Stevenson, N. Global wetland inventory–current status and future priorities. *Mar. Freshw. Res.* **1999**, *50*, 717–727. [CrossRef]
6. Dahl, T.E. *Status and Trends of Wetlands in Conterminous United States 1986 to 1997*; U.S. Fish and Wildlife Service, Branch of Habitat Assessment: Onalaska, WI, USA, 2000.
7. Dahl, T.E.; Watmough, M.D. Current approaches to wetland status and trends monitoring in prairie Canada and the continental United States of America. *Can. J. Remote Sens.* **2007**, *33*, 17–27. [CrossRef]
8. Creed, I.; Lane, C.; Serran, J.; Alexander, L.; Basu, N.; Calhoun, A.; Christensen, J.; Cohen, M.; Craft, C.; D'Amico, E.; et al. Enhancing protection for vulnerable waters. *Nat. Geosci.* **2017**, *10*, 809–815. [CrossRef]
9. Ozesmi, S.L.; Bauer, M.E. Satellite remote sensing of wetlands. *Wetl. Ecol. Manag.* **2002**, *10*, 381–402. [CrossRef]

10. Adam, E.; Mutanga, O.; Rugege, D. Multispectral and hyperspectral remote sensing for identification and mapping of wetland vegetation: A review. *Wetl. Ecol. Manag.* **2010**, *18*, 281–296. [CrossRef]

11. Hess, L.L.; Melack, J.M.; Filoso, S.; Wang, Y. Delineation of inundated area and vegetation along the Amazon floodplain with the SIR-C synthetic aperture radar. *IEEE Trans. Geosci. Remote Sens.* **1995**, *33*, 896–904. [CrossRef]

12. Wickham, J.; Stehman, S.; Smith, J.; Yang, L. Thematic accuracy of the 1992 National Land-Cover Data for the western United States. *Remote Sens. Environ.* **2004**, *91*, 452–468. [CrossRef]

13. Wright, C.; Gallant, A. Improved wetland remote sensing in Yellowstone National Park using classification trees to combine TM imagery and ancillary environmental data. *Remote Sens. Environ.* **2007**, *107*, 582–605. [CrossRef]

14. Bourgeau-Chavez, L.L.; Riordan, K.; Powell, R.B.; Miller, N.; Nowels, M. Improving wetland characterization with multi-sensor, multi-temporal SAR and optical/infrared data fusion. In *Advances in Geosciences and Remote Sensing*; InTechOpen Press: London, UK, 2009; pp. 679–708.

15. Finlayson, C.; Valk, A.G. Wetland classification and inventory: A summary. *Plant Ecol.* **1995**, *118*, 185–192. [CrossRef]

16. Guo, M.; Li, J.; Sheng, C.; Xu, J.; Wu, L. A review of wetland remote sensing. *Sensors* **2017**, *17*, 777. [CrossRef] [PubMed]

17. Mahdavi, S.; Salehi, B.; Granger, J.; Amani, M.; Brisco, B.; Huang, W. Remote sensing for wetland classification: A comprehensive review. *GISci. Remote Sens.* **2017**. [CrossRef]

18. Huguenin, R.L.; Karaska, M.A.; Van Blaricom, D.; Jensen, J.R. Subpixel classification of Bald Cypress and Tupelo Gum trees in Thematic Mapper imagery. *Photogramm. Eng. Remote Sens.* **1997**, *63*, 717–724.

19. Oki, K.; Oguma, H.; Sugita, M. Subpixel classification of alder trees using multitemporal Landsat Thematic Mapper imagery. *Photogramm. Eng. Remote Sens.* **2002**, *68*, 77–82.

20. Stankiewicz, K.; Dabrowska-Zielinska, K.; Gruszczynska, M.; Hoscilo, A. Mapping vegetation of a wetland ecosystem by fuzzy classification of optical and microwave satellite images supported by various ancillary data. *Remote Sens. Agric. Ecosyst. Hydrol.* **2003**, *4879*, 352–361.

21. Shanmugam, P.; Ahn, Y.H.; Sanjeevi, S. A comparison of the classification of wetland characteristics by linear spectral mixture modelling and traditional hard classifiers on multispectral remotely sensed imagery in Southern India. *Ecol. Model.* **2006**, *194*, 379–394. [CrossRef]

22. Fournier, R.A.; Grenier, M.; Lavoie, A.; Hélie, R. Towards a strategy to implement the Canadian wetland inventory using satellite remote sensing. *Can. J. Remote Sens.* **2007**, *33*, S1–S16. [CrossRef]

23. Grenier, M.; Labrecque, S.; Garneau, M.; Tremblay, A. Object-based classification of a SPOT-4 image for mapping wetlands in the context of greenhouse gases emissions: The case of the Eastmain region, Quebec, Canada. *Can. J. Remote Sens.* **2008**, *34*, S398–S413. [CrossRef]

24. Wang, J.; Lang, P.A. Detection of cypress canopies in the Florida Panhandle using subpixel analysis and GIS. *Remote Sens.* **2009**, *1*, 1028–1042. [CrossRef]

25. Frohn, R.; Autrey, B.; Lane, C.; Reif, M. Segmentation and object-oriented classification of wetlands in a karst Florida landscape using multi-season Landsat-7 ETM+ imagery. *Int. J. Remote Sens.* **2011**, *32*, 1471–1489. [CrossRef]

26. Powers, R.P.; Hay, G.J.; Chen, G. How wetland type and area differ through scale: A GEOBIA case study in Alberta's Boreal Plains. *Remote Sens. Environ.* **2011**, *117*, 135–145. [CrossRef]

27. Hird, J.; DeLancey, E.; McDermid, G.; Kariyeva, J. Google Earth Engine, open-access satellite data, and machine learning in support of large-area probabilistic wetland mapping. *Remote Sens.* **2017**, *9*, 1315. [CrossRef]

28. Ball, G.H.; Hall, D.J. *ISODATA, A Novel Method of Data Analysis and Pattern Classification*; DTIC Document; Stanford Research Inst. Menlo Park CA: Menlo Park, CA, USA, 1965.

29. Jain, A.; Dubes, R. *Algorithms for Clustering Data*; Prentice Hall: Englewood Cliffs, NJ, USA, 1988.

30. Jensen, J.R. *Introductory Digital Image Processing*, 3rd ed.; Prentice Hall: Upper Saddle River, NJ, USA, 2005.

31. Foody, G.M.; Campbell, N.A.; Trodd, N.M.; Wood, T.F. Derivation and applications of probabilistic measures of class membership from the maximum likelihood classification. *Photogramm. Eng. Remote Sens.* **1992**, *58*, 1335–1341.

32. Lek, S.; Guegan, J. Artificial neural networks as a tool in ecological modeling, an introduction. *Ecol. Model.* **1999**, *120*, 65–73. [CrossRef]

33. Dixon, B.; Candade, N. Multispectral landuse classification using neural networks and support vector machines: One or the other, or both? *Int. J. Remote Sens.* **2008**, *29*, 1185–1206. [CrossRef]

34. Mountrakis, M.; Im, J.; Ogole, C. Support vector machines in remote sensing: A review. *ISPRS J. Photogramm. Remote Sens.* **2011**, *66*, 247–259. [CrossRef]

35. Breiman, L. Bagging predictors. *Mach. Learn.* **1996**, *24*, 123–140. [CrossRef]

36. Quinlan, J.R. Learning decision tree classifiers. *ACM Comput. Surv. (CSUR)* **1996**, *28*, 71–72. [CrossRef]

37. Quinlan, J.R. Data Mining Tools see5 and c5. 0. 2004. Available online: http://www.rulequest.com/see5-info.html (accessed on 21 March 2018).

38. Kuhn, M.; Johnson, K. Classification trees and rule-based models. In *Applied Predictive Modeling*; Springer: New York, NY, USA, 2013; pp. 369–413.

39. Hansen, M.; Dubayah, R.; DeFries, R. Classification trees: An alternative to traditional land cover classifiers. *Int. J. Remote Sens.* **1996**, *17*, 1075–1081. [CrossRef]

40. DeFries, R.; Hansen, M.; Townshend, J.; Sohlberg, R. Global land cover classifications at 8 km spatial resolution: The use of training data derived from Landsat imagery in decision tree classifiers. *Int. J. Remote Sens.* **1998**, *19*, 3141–3168. [CrossRef]

41. Clark, L.; Pregibon, D.; Chambers, J.; Hastie, T. Tree-Based Models. In *Statistical Models in S*; Routledge: Abingdon, UK, 1992; pp. 377–419.

42. Baker, C.; Lawrence, R.; Montagne, C.; Patten, D. Mapping wetlands and riparian areas using Landsat ETM+ imagery and decision-tree-based models. *Wetlands* **2006**, *26*, 465–474. [CrossRef]

43. Friedl, M.A.; Brodley, C.E. Decision tree classification of land cover from remotely sensed data. *Remote Sens. Environ.* **1997**, *61*, 399–409. [CrossRef]

44. Li, J.; Chen, W. A rule-based method for mapping Canada's wetlands using optical, radar and DEM data. *Int. J. Remote Sens.* **2005**, *26*, 5051–5069. [CrossRef]

45. Sader, S.A.; Ahl, D.; Liou, W.-S. Accuracy of Landsat-TM and GIS rule-based methods for forest wetland classification in Maine. *Remote Sens. Environ.* **1995**, *53*, 133–144. [CrossRef]

46. Houhoulis, P.F.; Michener, W.K. Detecting wetland change: A rule-based approach using NWI and SPOT-XS data. *Photogramm. Eng. Remote Sens.* **2000**, *66*, 205–211.

47. Berhane, T.M.; Lane, C.R.; Wu, Q.; Anenkhonov, O.A.; Chepinoga, V.V.; Autrey, B.C.; Liu, H. Comparing pixel- and object-based approaches in effectively classifying wetland-dominated landscapes. *Remote Sens.* **2018**, *10*, 46. [CrossRef]

48. Kotsiantis, S. Combining bagging, boosting, rotation forest and random subspace methods. *Artif. Intell. Rev.* **2011**, *35*, 223–240. [CrossRef]

49. Belgiu, M.; Drăgut, L. Random forest in remote sensing: A review of applications and future directions. *ISPRS J. Photogramm. Remote Sens.* **2016**, *114*, 24–31. [CrossRef]

50. Breiman, L. Random forests. *Mach. Learn.* **2001**, *45*, 5–32.

51. Liaw, A.; Wiener, M. Classification and regression by randomForest. *R News* **2002**, *2*, 18–22.

52. Tian, S.; Zhang, X.; Tian, J.; Sun, Q.R. Random forest classification of wetland landcovers from multi-sensor data in the arid region of Xinjiang, China. *Remote Sens.* **2016**, *8*, 954. [CrossRef]

53. Corcoran, J.; Knight, J.; Gallant, A. Influence of multi-source and multi-temporal remotely sensed and ancillary data on the accuracy of random forest classification of wetlands in Northern Minnesota. *Remote Sens.* **2013**, *5*, 3212–3238. [CrossRef]

54. Van Beijma, S.; Comber, A.; Lamb, A. Random forest classification of salt marsh vegetation habitats using quad-polarimetric airborne SAR, elevation and optical RS data. *Remote Sens. Environ.* **2014**, *149*, 118–129.

55. Vanderhoof, M.K.; Alexander, L.C.; Todd, M.J. Temporal and spatial patterns of wetland extent influence variability of surface water connectivity in the Prairie Pothole Region, United States. *Landsc. Ecol.* **2016**, *31*, 805–824. [CrossRef]

56. DeVries, B.; Huang, C.; Lang, M.; Jones, J.; Hiang, W.; Creed, I.; Carroll, M. Automated quantification of surface water inundation in wetlands using optical satellite imagery. *Remote Sens.* **2017**, *9*, 807. [CrossRef]

57. Lane, C.R.; D'Amico, E. Calculating the ecosystem service of water storage in isolated wetlands using LiDAR in North Central Florida, USA. *Wetlands* **2010**, *30*, 967–977. [CrossRef]

58. Lane, C.R.; Autrey, B.C.; Jicha, T.; Lehto, L.; Elonen, C.; Seifert-Monson, L. Denitrification potential in geographically isolated wetlands of North Carolina and Florida, USA. *Wetlands* **2015**, *35*, 459–471.

59. Chalov, S.; Thorslund, J.; Kasimov, N.; Aybullatov, D.; Ilyicheva, E.; Karthe, D.; Kositsky, A.; Lychagin, M.; Nittrouer, J.; Pavlov, M.; et al. The Selenga River Delta: A geochemical barrier protecting Lake Baikal water. *Reg. Environ. Chang.* **2017**, *17*, 2039–2053. [CrossRef]

60. Lane, C.R.; Liu, H.; Autrey, B.C.; Anenkhonov, O.A.; Chepinoga, V.V.; Wu, Q. Improved wetland classification using eight-band high resolution satellite imagery and a hybrid approach. *Remote Sens.* **2014**, *6*, 12187–12216. [CrossRef]

61. Lane, C.R.; Anenkhonov, O.; Liu, H.; Autrey, B.C.; Chepinoga, V. Classification and inventory of freshwater wetlands and aquatic habitats in the Selenga River Delta of Lake Baikal, Russia, using high-resolution satellite imagery. *Wetl. Ecol. Manag.* **2015**, *23*, 195–214. [CrossRef]

62. Khazheeva, Z.; Tulokhonov, A.; Yao, R.; Hu, W. Seasonal and spatial distribution of heavy metals in the Selenga River Delta. *J. Geogr. Sci.* **2008**, *18*, 319–327. [CrossRef]

63. Brunello, A.J.; Molotov, V.C.; Dugherkhuu, B.; Goldman, C.; Khamaganova, E.; Strijhova, T.; Sigman, R. Lake Baikal Management Experience and Lessons Learned Brief. 2008. Available online: http://iwlearn.net/documents/10304 (accessed on 21 March 2018).

64. Garmaev, E.J.; Khristoforov, A.V. *Water Resources of the Rivers of the Lake Baikal Basin: Basics of Their Use and Protection*; Geo: Novosibirsk, Russia, 2010.

65. Potemkina, T. Hydrological–morphological zoning of the mouth zone of the Selenga River. *Water Resour.* **2004**, *31*, 11–16. [CrossRef]

66. Moore, M.; Hampton, S.; Izmest'eva, L.; Silow, E.; Peshkova, E.; Pavlov, B. Climate change and the world's "Sacred Sea" Lake Baikal, Siberia. *Bioscience* **2009**, *59*, 405–417. [CrossRef]

67. Thorslund, J.; Jarsjö, J.; Chalov, S.R.; Belozerova, E.V. Gold mining impact on riverine heavy metal transport in a sparsely monitored region: The upper Lake Baikal Basin case. *J. Environ. Monit.* **2012**, *14*, 2780–2792. [CrossRef] [PubMed]

68. Richards, J.A.; Jia, X. Feature reduction. In *Remote Sensing Digital Image Analysis*; Springer: Berlin/Heidelberg, Germany, 1999; pp. 239–257.

69. King, R. Land cover mapping principles: A return to interpretation fundamentals. *Int. J. Remote Sens.* **2002**, *23*, 3525–3545. [CrossRef]

70. Daniels, A.E. Incorporating domain knowledge and spatial relationships into land cover classifications: A rule-based approach. *Int. J. Remote Sens.* **2006**, *27*, 2949–2975. [CrossRef]

71. Tucker, C.J. Red and photographic infrared linear combinations for monitoring vegetation. *Remote Sens. Environ.* **1979**, *8*, 127–150. [CrossRef]

72. Gao, B. NDWI-A normalized difference water index for remote sensing of vegetation liquid water from space. *Remote Sens. Environ.* **1996**, *58*, 257–266. [CrossRef]

73. Wolf, A. *Using Worldview 2 Vis-NIR MSI Imagery to Support Land Mapping and Feature Extraction Using Normalized Difference Index Ratios*; Digital Globe: Longmont, CO, USA, 2010.

74. McFeeters, S.K. Using the normalized difference water index (NDWI) within a geographic information system to detect swimming pools for mosquito abatement: A practical approach. *Remote Sens.* **2013**, *5*, 3544–3561.

75. Parviainen, M.; Zimmermann, N.; Heikkinen, R.; Luoto, M. Using unclassified continuous remote sensing data to improve distribution models of red-listed plant species. *Biodivers. Conserv.* **2013**, *22*, 1731–1754. [CrossRef]

76. Sakamoto, T.; Van Nguyen, N.; Kotera, A.; Ohno, H.; Ishitsuka, N.; Yokozawa, M. Detecting temporal changes in the extent of annual flooding within the Cambodia and the Vietnamese Mekong Delta from MODIS time-series imagery. *Remote Sens. Environ.* **2007**, *3*, 295–313. [CrossRef]

77. Yamagata, Y.; Yasuoka, Y. Classification of wetland vegetation by texture analysis methods using ERS-1 and JERS-1 images. In Proceedings of the IEEE International Geoscience and Remote Sensing Symposium 1993 (IGARSS'93), Better Understanding of Earth Environment, Tokyo, Japan, 18–21 August 1993; pp. 1614–1616.

78. Franklin, S.E.; Peddle, D.R. Classification of SPOT HRV imagery and texture features. *Remote Sens.* **1990**, *11*, 551–556. [CrossRef]

79. Haralick, R.M.; Shanmugam, K.; Dinstein, I.H. Textural features for image classification. *IEEE Trans. Syst. Man Cybern.* **1973**, *3*, 610–621.

80. Gyninova, B.; Gyninova, A.; Balsanova, L. Genesis and evolution of soils in the Selenga river delta. *Mosc. Univ. Soil Sci. Bull.* **2008**, *63*, 171–177. [CrossRef]

81. Gyninova, A.; Korsunov, V. The soil cover of the Selenga delta area in the Baikal region. *Eur. Soil Sci.* **2006**, *39*, 243–250. [CrossRef]

82. Liu, H.; Wang, L.; Sherman, D.; Gao, Y.; Wu, Q. An object-based conceptual framework and computational method for representing and analyzing coastal morphological changes. *Int. J. Geogr. Inf. Sci.* **2010**, *24*, 1015–1041. [CrossRef]

83. Frick, A.; Steffenhagen, P.; Zerbe, S.; Timmermann, T.; Schulz, K. Monitoring of the vegetation composition in rewetted peatland with iterative decision tree classification of satellite imagery. *Photogramm. Fernerkund. Geoinf.* **2011**, *2011*, 109–122.

84. Hansen, M.; Defries, R.; Townshend, J.; Sohlberg, R. Global land cover classification at 1 km spatial resolution using a classification tree approach. *Int. J. Remote Sens.* **2000**, *21*, 1331–1364. [CrossRef]

85. DeFries, R.; Chan, J.C.-W. Multiple criteria for evaluating machine learning algorithms for land cover classification from satellite data. *Remote Sens. Environ.* **2000**, *74*, 503–515. [CrossRef]

86. Simard, M.; Grandi, G.D.; Saatchi, S.; Mayaux, P. Mapping tropical coastal vegetation using JERS-1 and ERS-1 radar data with a decision tree classifier. *Int. J. Remote Sens.* **2002**, *23*, 1461–1474. [CrossRef]

87. Kearns, M.; Mansour, Y.; Ng, A.Y. An information-theoretic analysis of hard and soft assignment methods for clustering. In *Learning in Graphical Models*; Springer: Berlin/Heidelberg, Germany, 1998; Volume 89, pp. 495–520.

88. Lakkaraju, H.; Bach, S.; Leskovec, J. Interpretable decision sets: A joint framework for description and prediction. In Proceedings of the 22nd ACM SIGKDD International Conference on Knowledge Discovery and Data Mining (KDD'16), San Francisco, CA, USA, 13–17 August 2016.

89. Chan, J.; Paelinckx, D. Evaluation of random forest and adaboost tree-based ensemble classification and spectral band selection for ecotope mapping using airborne hyperspectral imagery. *Remote Sens. Environ.* **2008**, *112*, 2999–3011. [CrossRef]

90. Foody, G. Classification accuracy comparison: Hypothesis tests and the use of confidence intervals in evaluations of difference, equivalence and non-inferiority. *Remote Sens.* **2009**, *113*, 1658–1663. [CrossRef]

91. Rodriguez-Galiano, V.; Ghimire, B.; Rogan, J.; Chica-Olmo, M.; Rigol-Sanchez, J. An assessment of the effectiveness of a random forest classifier for land-cover classification. *ISPRS J. Photogramm. Remote Sens.* **2012**, *67*, 93–104. [CrossRef]

92. Prasad, A.M.; Iverswon, L.R.; Liaw, A. Newer classification and regression tree techniques: Bagging and random forests for ecological prediction. *Ecosystems* **2006**, *9*, 181–199.

93. Stumpf, A.; Kerle, N. Object-oriented mapping of landslides using random forests. *Remote Sens. Environ.* **2011**, *115*, 2564–2577. [CrossRef]

94. Millard, K.; Richardson, M. On the importance of training data sample selection in random forest image classification: A case study in peatland ecosystem mapping. *Remote Sens.* **2015**, *7*, 8489–8515. [CrossRef]

95. Wu, Q. GIS and remote sensing applications in wetland mapping and monitoring. In *Comprehensive Geographic Information Systems*; Huang, B., Ed.; Elsevier: Oxford, UK, 2018; pp. 140–157.

96. Dronova, J. Object-based image analysis in wetland research: A review. *Remote Sens.* **2015**, *7*, 6380–6413. [CrossRef]

97. Dubeau, P.; King, D.; Unbushe, D.; Rebelo, L. Mapping the Dabus wetlands, Ethiopia, using random forest classification of Landsat, PALSAR and topographic data. *Remote Sens.* **2017**, *9*, 1056. [CrossRef]

98. Wuest, B.; Zhang, Y. Region based segmentation of Quickbird multispectral imagery through bands ratios and fuzzy comparison. *ISPRS J. Photogramm. Remote Sens.* **2009**, *64*, 55–64. [CrossRef]

remote sensing

MDPI

Article

A New Method for Mapping Aquatic Vegetation Especially Underwater Vegetation in Lake Ulansuhai Using GF-1 Satellite Data

Qi Chen, Ruihong Yu *, Yanling Hao *, Linhui Wu, Wenxing Zhang, Qi Zhang and Xunan Bu

Inner Mongolia Key Laboratory of River and Lake Ecology & Ministry of Education Key Laboratory of Ecology and Resource Use of the Mongolian Plateau, School of Ecology and Environment, Inner Mongolia University, Hohhot 010021, China; qichen@mail.imu.edu.cn (Q.C.); wulinhui@imu.edu.cn (L.W.); 31514041@mail.imu.edu.cn (W.Z.); 31614024@mail.imu.edu.cn (Q.Z.); 31614033@mail.imu.edu.cn (X.B.)
* Correspondence: rhyu@imu.edu.cn (R.Y.); haoyl@imu.edu.cn (Y.H.)

Received: 18 July 2018; Accepted: 12 August 2018; Published: 14 August 2018

Abstract: It is difficult to accurately identify and extract bodies of water and underwater vegetation from satellite images using conventional vegetation indices, as the strong absorption of water weakens the spectral feature of high near-infrared (NIR) reflected by underwater vegetation in shallow lakes. This study used the shallow Lake Ulansuhai in the semi-arid region of China as a research site, and proposes a new concave–convex decision function to detect submerged aquatic vegetation (SAV) and identify bodies of water using Gao Fen 1 (GF-1) multi-spectral satellite images with a resolution of 16 meters acquired in July and August 2015. At the same time, emergent vegetation, "Huangtai algae bloom", and SAV were classified simultaneously by a decision tree method. Through investigation and verification by field samples, classification accuracy in July and August was 92.17% and 91.79%, respectively, demonstrating that GF-1 data with four-day short revisit period and high spatial resolution can meet the standards of accuracy required by aquatic vegetation extraction. The results indicated that the concave–convex decision function is superior to traditional classification methods in distinguishing water and SAV, thus significantly improving SAV classification accuracy. The concave–convex decision function can be applied to waters with SAV coverage greater than 40% above 0.3 m and SAV coverage 40% above 0.1 m under 1.5 m transparency, which can provide new methods for the accurate extraction of SAV in other regions.

Keywords: aquatic vegetation; concave–convex decision function; remote sensing extraction; GF-1 satellite; Lake Ulansuhai; China

---

1. Introduction

Aquatic vegetation plays an important role in the regulation of lake ecosystems, but in recent years, lake water quality has continuously deteriorated in semi-arid areas. The declining water quality is marked with severe eutrophication, frequent algal blooms, shrinking areas with aquatic vegetation, and even extinction of some vegetation [1]. To better provide early warnings of potential algal bloom outbreaks and accomplish dynamic monitoring of aquatic vegetation, rapid, large-scale, and regular monitoring of aquatic vegetation via remote sensing is an indispensable tool [2,3]. In the early years of remote sensing technology, aerial images were utilized to monitor aquatic vegetation [4,5]. As remote sensing technologies evolved, moderate-resolution imaging spectroradiometer (MODIS) satellite images with low resolution and high frequency [6,7]; Landsat thematic mapper (TM), enhanced thematic mapper plus (ETM+); and Huangjing-1A/B (HJ-1A/B) images with medium resolution [8–10]; as well as QuickBird, IKONOS, and other high-resolution images became available [11–13]. Meanwhile, many extraction methods for aquatic vegetation classification have been developed, such as decision tree classification [14], supervised classification [15], and unsupervised classification [16].

The decision tree classification method is especially helpful, and is widely used in aquatic vegetation classification [17–20]. In decision tree classification research that simultaneously extracts multiple types of aquatic vegetation, normalized difference vegetation index (NDVI) and normalized difference water index (NDWI) values have commonly been used as the classification variables for submerged aquatic vegetation (SAV) or other aquatic vegetation [17,18], and the simple ratio (SR) and ratio vegetation index (RVI) have also been used [19]. However, because the threshold range of these indices for identifying SAV overlaps with the threshold range of these indices for water, SAV and water are commonly confused during extraction and identification [21]. Therefore, in multispectral classification studies, conventional vegetation indices are only able to extract SAV with a high reflectance in the near-infrared (NIR) band. While the spectral signal of plants that grow underwater is easily inhibited by the strong absorption of the surrounding water, the high reflectance in the NIR band is weakened and even disappears, leading to decreased accuracy. To accurately identify and extract SAV from remote sensing data, classification based on the use of auxiliary information such as transparency was proposed, and provided a relatively good classification result [22,23]. However, this method is labor-intensive, as it requires simultaneous investigation of various kinds of auxiliary information. In addition, high-resolution images or hyperspectral data are widely used to achieve a more accurate classification result for SAV [24–26], and the calibration method of the spectral curve also contributes to the effective extraction of SAV [27,28]. The efficient and accurate extraction of SAV in research has so far mainly concentrated on the hyperspectral field, while there are few simple and effective methods for the simultaneous extraction of multiple kinds of aquatic vegetation (as well as vegetation covered by water) based on multispectral data on a large scale.

The multispectral remote sensing Gao Fen 1 (GF-1) satellite carries a 2 m panchromatic camera and an 8 m multispectral camera, as well as four multispectral cameras with a resolution of 16 m. It thus produces data with higher resolution than MODIS, TM, and HJ-1A/B. With a combined large detection width of 800 km and a relatively short revisit period of four days, these parameters are important in obtaining detailed monitoring of vegetation growth. While GF-1 is used mostly in terrestrial vegetation monitoring (e.g., forest land, grassland, crops, etc.) [29–31], it has fewer applications in monitoring aquatic vegetation.

The study aimed to test the suitability of GF-1 data for the detection and mapping of SAV in small lakes by meeting the following two objectives: (1) developing a novel decision function to efficiently distinguish SAV from water; and (2) simultaneously classifying emergent vegetation and Huangtai algae concomitant with SAV using a decision tree model. The study was performed at Lake Ulansuhai, China, a shallow weed-type lake in an arid area. This research goes beyond single object extraction, using a simple and effective method to simultaneously extract emergent vegetation, SAV, and Huangtai algae information. This work provides a potential complete and effective method for the long-term monitoring of aquatic vegetation via the effective classification of submerged vegetation and water bodies.

2. Materials and Methods

2.1. Study Area

Lake Ulansuhai is the largest freshwater lake in the Yellow River Basin, and the only drainage area in the Hetao irrigation region. Located in Bayannur, Inner Mongolia, Lake Ulansuhai has a longitude from $108°43'$–$108°57'$E and a latitude from $40°36'$–$41°03'$N. The region lies in a temperate continental climate and has alternating seasons with a multiyear average precipitation of 221.1 mm and a multiyear average evaporation of 2382.1 mm. The lake depth ranges from 0.5–2.5 m, and the storage capacity of the lake is 0.32 billion m^3. Lake Ulansuhai is classified as a severely eutrophic weed-type lake [32]. The aquatic vegetation in Lake Ulansuhai could be generally classified as emergent vegetation, SAV, and Huangtai algae. Huangtai algae are composed of multiple filamentous algae of chlorophytes Zygophyceae, Zygnematales, and Zygnemataceae, and mainly contain *Spirogyra*, *Zygnema*, and *Mongeotia* algae [33]. Dominant species of aquatic vegetation in Lake Ulansuhai are shown in Figure 1.

Figure 1. Dominant species of aquatic vegetation in Lake Ulansuhai. SAV: submerged aquatic vegetation.

2.2. Remote Sensing Data and Processing

The GF-1 satellite carries a 2-m panchromatic camera, an 8-m multispectral camera, and four 16-m wide field view (WFV) cameras, and was launched by China on 26 April 2013. To classify aquatic vegetation, GF-1 WFV images with a resolution of 16 m were chosen, including three visible light bands (blue (band 1), green (band 2), and red (band 3)), and one NIR band (band 4), which are similar to the first four bands of Landsat TM images. With the four multispectral cameras combined, a swath width of 800 km was achieved. The specifications of GF-1 WFV cameras are shown in Table 1. Two images that showed abundant vegetation information from Lake Ulansuhai on 2 July and 8 August in 2015 were selected as the classification data source (because the image on 8 August 2015 contained some cloud, the cloudy part was removed). The Environment for Visualizing Images (ENVI) was used for image pre-processing of ortho-rectification, radiometric calibration, and atmospheric correction. The Fast Line-of-sight Atmospheric Analysis of Spectral Hypercubes (FLAASH) algorithm was adopted for atmospheric correction. The FLAASH model has been used as an effective method for the atmospheric correction of GF-1 images, and can provide accurate surface reflectance [34,35]. Landsat8 Operational Land Imager (OLI) data was used as the reference image for geometric correction, with correction error controlled within 0.5 pixels. The GF-1 data were provided by the China Centre for Resources Satellite Data and Application.

Table 1. Characterization of Gao Fen 1 (GF-1) wide field view (WFV) cameras.

Sensor	Band	Spectral Range (μm)	Band Type	Spatial Resolution (m)	Swath Width (km)	Revisit Period (days)	Orbit Altitude (km)
WFV (1–4)	1	0.45–0.52	Blue	16	800	4	645
	2	0.52–0.59	Green				
	3	0.63–0.69	Red				
	4	0.77–0.89	NIR				

2.3. Acquisition of Field Data

Approximately at the acquisition time of the GF-1 images, a simultaneous field investigation was carried out from 2–5 July and 8–10 August 2015, totaling 146 investigation points, as shown in Figure 2. In July and August 2015, the numbers of emergent vegetation sample points were 23 and 21, the numbers of SAV sample points were 32 and 25, and the numbers of Huangtai algae sample points were 25 and 20, respectively. The sampling locations were set according to the aquatic vegetation distribution, and were chosen such that the investigation covered the whole lake. The location coordinates and type of vegetation at each point were recorded. The sample area for all investigation points was over 64 m × 64 m (or equivalent to four pixels of GF-1 data). To further explore the spectral curve characteristics of SAV and water, the reflectance of SAV and water were measured with an Analytica Spectra Devices, Inc. (ASD) FieldSpec® Handheld2™ Spectroradiometer in July 2016. The spectral curve changes of SAV with depth under different conditions were also measured in June 2018 unsing ASD FieldSpec® 4 spectroradiometer. The wavelength response range of FieldSpec® Handheld2™ and FieldSpec® 4 spectroradiometer is 325–1075 nm and 350–2500 nm, respectively. Because of the presence of noise in the signal, we used only the range 350–900 nm.

2.4. Methods

The aquatic vegetation in Lake Ulansuhai was divided into three types, based on field monitoring: emergent vegetation, SAV, and Huangtai algae. Because of the flowering of vegetation in August, WFV1 image from GF-1 on 8 August 2015 was selected to analyze the spectral characteristics of different classes. ENVI 5.1 was used to plot the average spectral signature for different classes by combining endmember sample points at different locations for each class.

2.4.1. Identification and Detection of Land and Emergent Vegetation

According to Figure 3, the land had a very high reflectance in the third band, meaning that band 3 could be used to identify land. Because of the chlorophyll content of emergent vegetation, it had strong absorption in blue and red light. Therefore, absorption valleys appeared in bands 1 and 3. As the emergent vegetation absorbed less green light, a small reflectance peak appeared in the green band. In the NIR band, the different refractive indices of the cell wall and lacuna inside the leaf caused multiple reflections, forming a high-reflection region [36]. The reflectance of emergent vegetation in band 4 was higher than the reflectance of other vegetation, and because the reflectance difference in bands 2 and 3 was also relatively high, band 4 or the combination of bands 2 and 3 could be used to extract the presence of emergent vegetation.

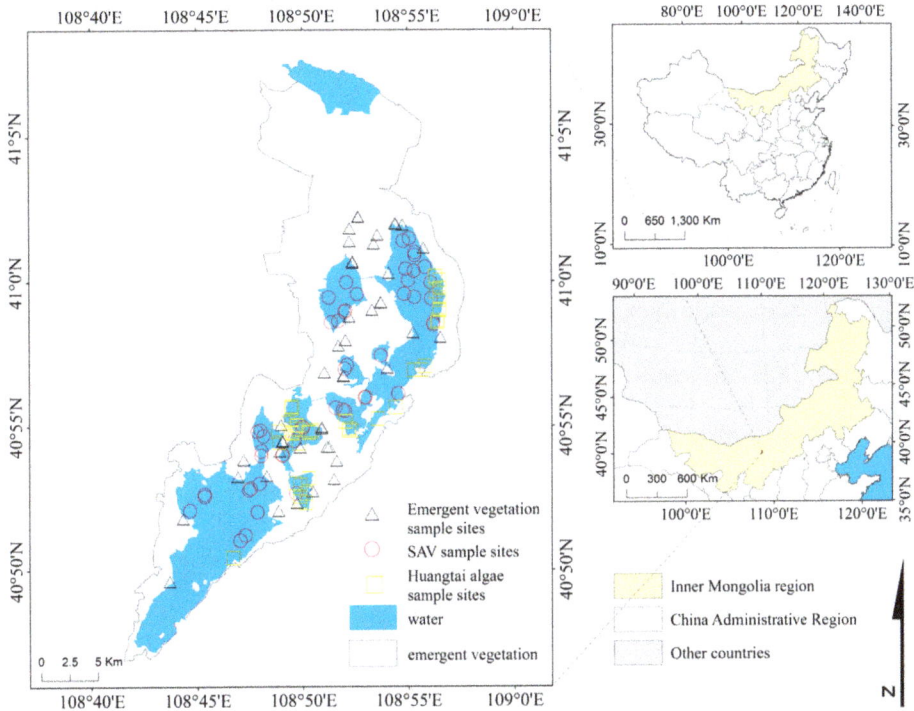

Figure 2. Aquatic vegetation samples in Lake Ulansuhai.

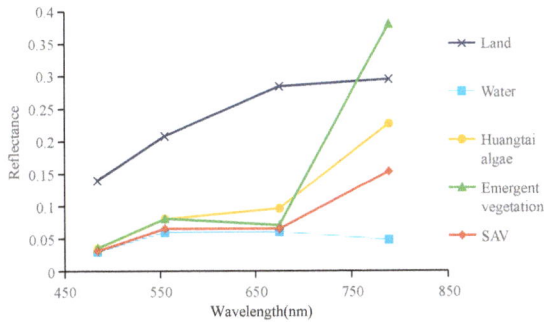

Figure 3. Spectral curves of different classes.

2.4.2. Identification and Detection of Huangtai Algae

Figure 3 shows that the spectral curve of Huangtai algae showed a high reflectance typical of vegetation in NIR. Therefore, NDVI could be used to distinguish Huangtai algae from the water body. The spectral curve of Huangtai algae with low density was observed to be similar to that of SAV. Huangtai algae has a yellowish color, which leads to increased reflectance in the red-light band (band 3) [33]. Huangtai algae had a relatively high reflectance in both bands 2 and 3, but SAV did not show a high reflectance in band 3 due to absorption at this wavelength, meaning that Huangtai algae and SAV could be distinguished from each other using B3–B2, calculated from their difference in bands 2 and 3.

2.4.3. Identification and Detection of Water and SAV

The reflectance of water and SAV fluctuates within a certain range around the average spectral curve. Because of the effects of water, the reflectance in the NIR and the red-light bands of aquatic vegetation growing underwater is weakened [37]. A rapid decrease in reflectance in the NIR band was observed as the aquatic vegetation was covered by water, but the absorption of red light was less. When SAV growth reached 43–51 cm below clear water, Cho et al. measured that the NDVI value was close to zero using multiple sensors because the NIR reflectance was completely weakened to the value of red-light reflectance [38].

Figure 4a,b show the field-measured submerged vegetation and water spectral curves. The macrophytes above the water surface showed typical spectral characteristics of green vegetation, with reflection valleys at approximately 675 nm, a sharp reflectance increase at approximately 700 nm, and strong reflection in the NIR band (770–890 nm). The water also showed a reflection peak near 700 nm, after which the reflectance decreased because of the strong absorption of the water body, resulting in a low reflectance in the NIR band. Therefore, it was easy to divide the macrophytes above the water surface from the water body by judging whether there was a high reflectance in the NIR band. However, because of the impact of the water body, the reflectance of macrophytes below the water surface in the NIR band was greatly reduced and did not show a high reflectance, making them difficult to distinguish from the water body.

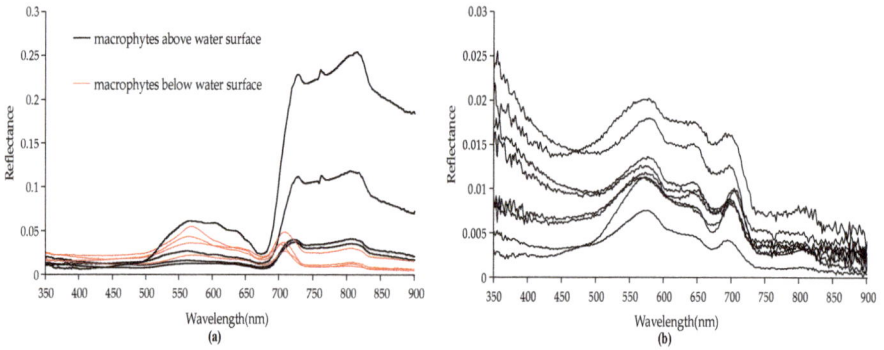

Figure 4. The field-measured spectral curves of (**a**) submerged vegetation and (**b**) water.

Based on the spectral response function of WFV1, the convolution method was applied to the ASD spectral field data to calculate the equivalent surface reflectance of submerged vegetation and water in four bands. We simulated GF-1 satellite data to further analyze the spectral characteristics of submerged vegetation and water. The equation is as follows:

$$< R_i >= \frac{\int_{\lambda_1}^{\lambda_2} S_i(\lambda) R(\lambda) d\lambda}{\int_{\lambda_1}^{\lambda_2} S_i(\lambda) d\lambda} \tag{1}$$

where R_i denotes the equivalent surface reflectance of the ith band of WFV1, λ_1 to λ_2 is the spectral wavelength range of the ith band of WFV1, $R(\lambda)$ is the corresponding reflectance at wavelength λ, and $S_i(\lambda)$ is the corresponding response value of the spectral response function of the ith band of WFV1 at wavelength λ.

The equivalent reflectance of submerged vegetation and the water body are shown in Figure 5a–c. The macrophytes above the water surface showed high reflectance in the NIR band and could be easily distinguished from the water body (Figure 5a). Figure 5b shows that some macrophytes growing below the water surface did not show high reflectance in the NIR band, and the reflectance of band 4 was lower than the reflectance of band 3, making its spectral characteristic curves similar to the spectral characteristic curves of water, increasing the difficulty of identifying the macrophytes using NDVI. The water spectral curve inflection point of band 3 remained convex, while the SAV curve still had an absorption valley in band 3 with the curve around this inflection point either concave or convex with small convexity, which is consistent with the experimental results obtained by Cho et al. [39]. To distinguish between SAV and water, a concave–convex decision function was constructed, expressed as follows:

$$F = (B4 - B3)/0.114 - (B3 - B2)/0.12 \tag{2}$$

in which F denotes the concave–convex decision function. The value 0.114 denotes the difference between the central wavelength of band 4 and band 3; 0.12 denotes the difference between the central wavelength of band 3 and band 2. $(B4 - B3)/0.114$ denotes the slope of the spectral curve between bands 3 and 4, $k1$; and $(B3 - B2)/0.12$ denotes the slope of the spectral curve between bands 2 and 3, $k2$.

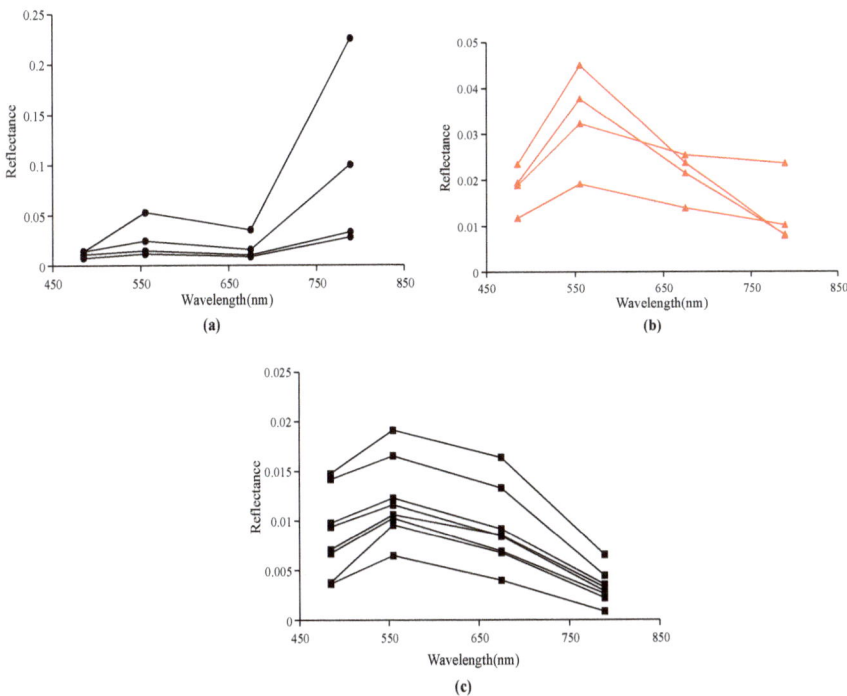

Figure 5. Equivalent surface reflectance of submerged vegetation and water. Equivalent reflectance of macrophytes (**a**) above and (**b**) below the water surface, respectively. (**c**) The equivalent reflectance of water.

Equation (2) denotes $k1 - k2$. When the curve around the inflection point is concave, $(k1 - k2) > 0$. When the curve around the inflection point is convex, $(k1 - k2) < 0$. Thus, a concave shape yields positive values, while a convex shape yields negative values.

Spectral characteristic curves of water and aquatic vegetation covered by water on GF-1 WFV images are shown in Figure 6. In this figure, the spectral curve of water had a convex shape at the inflection point of band 3, and the spectral curve of SAV had a concave shape at the inflection point of band 3. If we assume the spectral characteristic curve functions of water and SAV are $A(X)$ and $B(X)$, respectively, $k1$, $k2$, $k1 - k2$, $|k1 - k2|$, and obtuse angle (α) at the inflection point can be calculated as shown in Table 2. Table 2 shows that $k1 - k2$ of the spectral curve of water was negative, while $k1 - k2$ of the SAV spectral curve was positive. As the relative concavity or convexity of the spectral curve at the inflection point in band 3 increased, the obtuse angle decreased, and $|k1 - k2|$ increased. In this case, $|k1 - k2|$ is an appropriate metric to quantify the concavity and convexity of the different spectra.

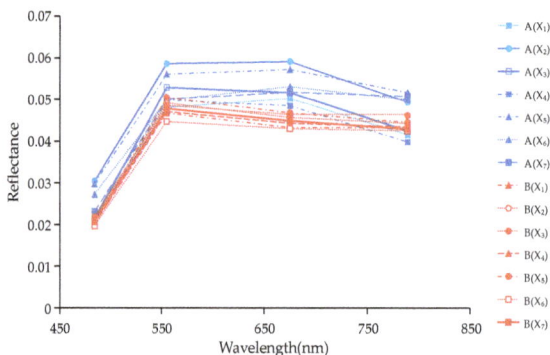

Figure 6. Spectral curves of water and macrophytes underwater. Blue curves denote spectral curves of water, and red curves denote spectral curves of SAV. $A(X)$ and $B(X)$ denote the spectral characteristic curve functions of water and SAV, respectively.

Table 2. Calculated values of the spectral curve function for water and macrophytes underwater.

	$k1$	$k2$	$k1 - k2$	$|k1 - k2|$	Obtuse Angle α
$A(X1)$	−0.0763	0.0175	−0.0938	0.0938	174.6333
$A(X2)$	−0.0851	0.0042	−0.0893	0.0893	174.8978
$A(X3)$	−0.0816	−0.0100	−0.0716	0.0716	175.9091
$A(X4)$	−0.0763	−0.0117	−0.0646	0.0646	176.3043
$A(X5)$	−0.0482	0.0092	−0.0574	0.0574	176.7127
$A(X6)$	−0.0272	0.0267	−0.0539	0.0539	176.9148
$A(X7)$	−0.0088	0.0150	−0.0238	0.0238	178.6380
$B(X1)$	−0.0018	−0.0283	0.0266	0.0266	178.4776
$B(X2)$	−0.0123	−0.0292	0.0169	0.0169	179.0329
$B(X3)$	−0.0018	−0.0183	0.0166	0.0166	179.0502
$B(X4)$	−0.0079	−0.0233	0.0154	0.0154	179.1157
$B(X5)$	−0.0202	−0.0308	0.0107	0.0107	179.3898
$B(X6)$	−0.0044	−0.0142	0.0098	0.0098	179.4397
$B(X7)$	−0.0175	−0.0242	0.0066	0.0066	179.6207

Changes in transparency, depth, and coverage of SAV affect the reflectance of SAV. In order to further explore the transferability of the concave–convex decision function, we studied the effect of transparency, SAV depth, and coverage on the spectral curves of SAV. We conducted experiments with two different transparencies (0.6 m and 1.5 m), and we also set experiments on SAV with different coverage. SAV coverage ranged from 40% to 100%, and the SAV depth below the water surface ranged from 0 m to 1.3 m. The reflectance spectra of SAV was integrated to the four spectral bands of GF-1 using Equation (1).

2.4.4. Establishment of the Classification Tree Model

We propose decision variables based on the above analysis, namely, the concave–convex decision function, the single band, and combinations of multiple-bands. These were incorporated in the construction of a decision tree model for aquatic vegetation classification, as shown in Figure 7, where DV denotes decision variable, and a, b, c, d, e, and f denote the optimum threshold values of the decision variables.

Threshold values in the decision tree were repeatedly adjusted and modified according to 50% of the field survey sample points to obtain the optimum threshold values (the other 50% of field sample points were used for validation).

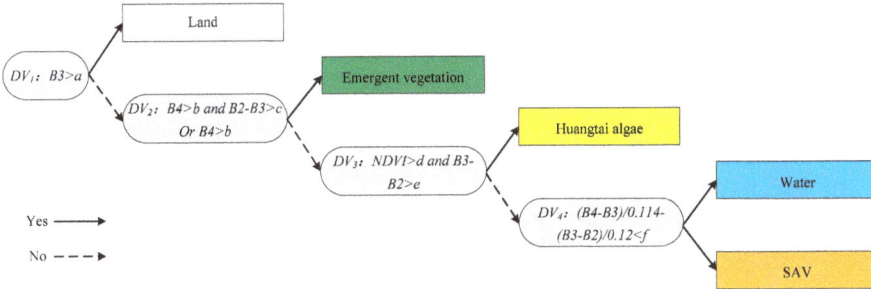

Figure 7. Classification tree model of aquatic vegetation. DV denotes a decision variable, and a, b, c, d, e, and f denote the optimum threshold values of the decision variables. NDVI: normalized difference vegetation index.

3. Results

3.1. Separability of Spectral Characteristic Variables

NDVI can generally differentiate between water and vegetation, but it is unable to efficiently extract aquatic vegetation underwater because of the interference from the body of water. The image of Lake Ulansuhai in August of 2015 was used to calculate the frequency distribution of different objects in the areas of interest. Figure 8a shows that water and SAV had a large overlapping area when classified using NDVI, which led to a decrease in classification accuracy as water and SAV were confused. However, when classified using the concave–convex decision function, water and SAV could be distinguished well (Figure 8b), indicating that the concave–convex decision function was better than NDVI in the accurate classification of water and SAV. Figure 8c shows that emergent vegetation was distinguishable from other vegetation using band B4. Huangtai algae and SAV could be differentiated using a combination of bands B3 and B2 (Figure 8d).

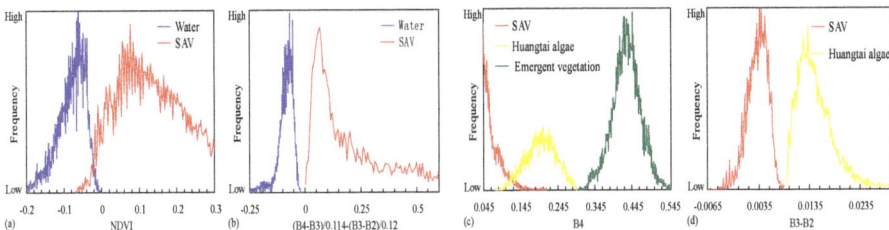

Figure 8. Statistical frequency distribution of different classes. The statistical frequency distribution of water and SAV using (**a**) NDVI and the (**b**) concave–convex decision function extraction method. The statistical frequency distribution of three species of vegetation is shown in (**c,d**).

3.2. Classification Results and Validation

Information about aquatic vegetation was extracted from satellite images in July and August of 2015 using the aquatic vegetation classification tree model. The spatial distribution of all vegetation types was obtained as shown in Figure 9.

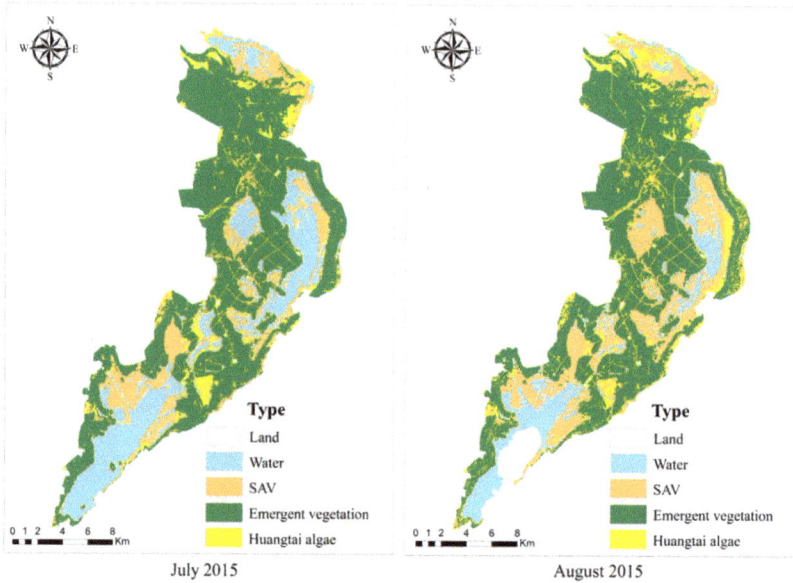

July 2015   August 2015

Figure 9. Aquatic vegetation classification results of remote sensing images in July and August 2015.

The confusion matrix and kappa coefficient [40,41] are commonly used methods for the evaluation of the accuracy of vegetation classification. Each column in the confusion matrix represents the ground-truthed classification of one class, and each value in the column is equal to the number of real pixels of the ground surface that were classified in different categories in the classification graph [42].

The confusion matrix based on 50% of the sample data from both the July and August 2015 image classification is shown in Table 3, showing that the overall accuracy of the two classifications was 92.17% and 91.79%, respectively, and the kappa coefficients were 0.8995 and 0.8935, respectively. These results indicate a relatively good classification performance, which provides a theoretical basis for dynamic monitoring of aquatic vegetation in Lake Ulansuhai.

Table 3. Classification accuracy test. SAV: submerged aquatic vegetation.

		Real Value											
		Land		Water		SAV		Emergent Vegetation		Huangtai Algae		Total	
	Month	07	08	07	08	07	08	07	08	07	08	07	08
	Land	19	15	0	0	0	0	0	0	1	0	20	15
Classification Value	Water	2	0	31	33	1	1	0	0	0	0	34	34
	SAV	0	0	2	2	43	61	4	3	2	1	51	67
	Emergent Vegetation	0	0	0	0	0	0	55	39	1	2	56	41

Table 3. *Cont.*

	Real Value										
	Land		Water		SAV		Emergent Vegetation		Huangtai Algae		Total	
Huangtai	2	5	0	0	2	3	0	0	52	42	56	50
Total	23	20	33	35	46	65	59	42	56	45	217	207
Producer Accuracy (%)	82.61	75.00	93.94	94.29	93.48	93.85	93.22	92.86	92.86	93.33		
User Accuracy (%)	95.00	100.00	91.18	97.06	84.31	91.04	98.21	95.12	92.86	84.00		
Kappa Coefficient	0.8995	0.8935										
Overall accuracy	92.17%	91.79%										

The producer and user accuracies of SAV in August 2015 were 93.85% and 91.04%, respectively. The producer and user accuracies of Huangtai algae were 93.33% and 84.00%, respectively. NDVI was also used to distinguish between water and SAV, resulting in SAV producer and user accuracies of 69.23% and 84.91%, respectively, and in SAV test samples, 26.15% of SAV was classified as water. Clearly, the method proposed in this study greatly improved the classification accuracy.

3.3. SAV Spectral Curve Changes with Depth under Different Transparency and Coverage

Figure 10 shows the variation of the SAV reflectance with depth below the water surface under different transparencies in the four bands of GF-1. The vegetation coverage was 100% and the water depth varied from 0 m to 0.5 m and 1.3 m. The NIR reflectance continuously decreased with increasing water depth. With 0.6 m transparency, the typical NIR high reflectance in the vegetation spectra was preserved up to 0.1 m, but was not visible for depths greater than 0.3 m. In a water body with a transparency of 1.5 m, the NIR reflectance decreased by 85.82% from 0 m to 0.3 m, but the NIR reflectance was still higher than that in the red band.

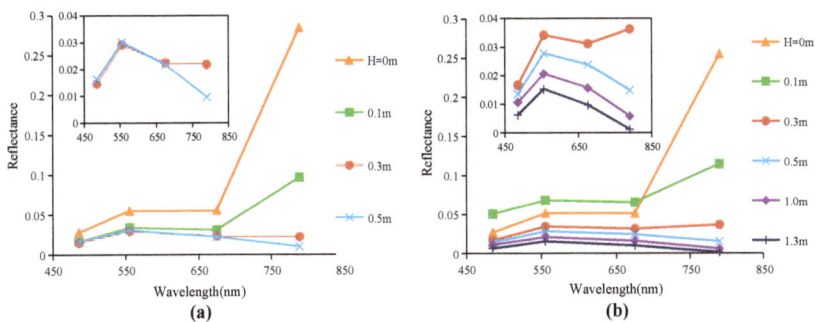

Figure 10. Remote sensing reflectance of SAV with 100% vegetation coverage at varying depth below water surface under different transparencies: (**a**) transparency = 0.6 m; and (**b**) transparency = 1.5 m. H represents different SAV depth.

Figure 11 shows the effects of vegetation coverage and depth on the SAV reflectance. When the SAV coverage was 80% and 60%, the NIR high peak could be detected above 0.1 m, and was not visible at 0.3 m. The NIR reflectance of SAV with 80% and 60% coverage decreased by 91.61% and 92.51%

from 0 m to 0.3 m, respectively. When SAV coverage was 40%, the NIR reflectance of SAV from 0 m to 0.3 m decreased by 92.29%, and the NIR did not show high reflection characteristics.

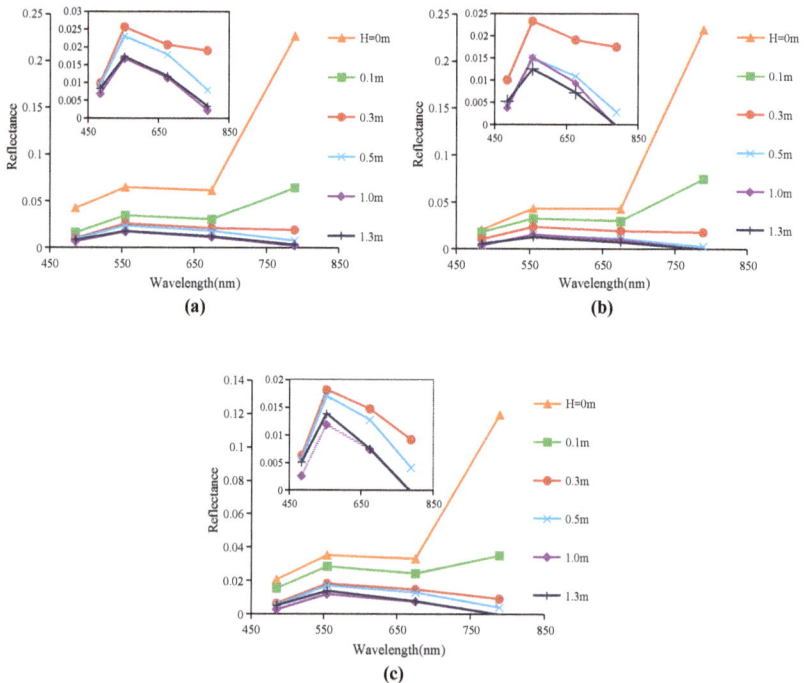

Figure 11. SAV remote sensing reflectance at different depths below the water surface under 1.5 m transparency: (**a**) SAV coverage = 80%; (**b**) SAV coverage = 60%; and (**c**) SAV coverage = 40%.

4. Discussion

In this study, a decision tree classification was developed for classifying GF-1 imagery to extract the aquatic vegetation in Lake Ulansuhai. The use of this decision tree achieved high classification accuracy in two GF-1 images. However, there are still certain limitations to the application of this method. Huangtai algae started growing in May, and it was difficult to detect it at this time. In addition, SAV turned yellow after September, and its reflectance no longer showed any significant difference from the Huangtai algae in band 3. Therefore, this decision tree method is only applicable to the flowering season of aquatic vegetation in summer. Changes in the weather condition also affected the classification results, and thus the method is applicable to clear and cloudless weather (as applies to most satellite surveys). Most confusion of this method occurred in the extraction of land. This was due to the existence of mixed pixels, and some land edges of islands in the lake were classified as Huangtai algae. In the verification of the classification results, the producer accuracy of land in July and August was 82.61% and 75%, respectively. The water bodies achieved high classification accuracy in July and August, and their producer accuracy was 93.94% and 94.29%, respectively. However, a small part of the water was still classified as SAV because, in some channels, water had similar spectral characteristics to that of nearby vegetation due to limits in image resolution, indicating that it might be difficult to accurately extract channel water for remote sensing products at this resolution. In July, only 2.17% of SAV was classified as water, which was mainly due to the mixed pixels of SAV and water. Only 1.54% of the SAV was classified as water in August, which was mainly caused by low vegetation coverage and water transparency. The error of the emergent vegetation was mainly present in its incorrect division into SAV, which was largely due to the sparseness of the reeds in some areas and mixed pixels with the water and SAV. Huangtai algae achieved high producer accuracy of 92.86% and 93.33% in July and August, respectively. In July, 3.57% of Huangtai algae was mistakenly classified as SAV. A total of 2.22% of Huangtai algae was misclassified as SAV in August. This may have been caused by the sparseness of Huangtai algae and mixture in the same pixel with SAV.

This study demonstrates that there is a sufficient difference in the spectral concavity and convexity between macrophytes below the water surface and water, and the concave–convex decision function could efficiently identify and detect the aquatic vegetation below the water surface. The spectral reflectance curves of submerged vegetation areas and non-submerged vegetation areas also exhibited similar concave–convex characteristics in Lake Pontchartrain [39]. However, the spectral signal of SAV is affected by various factors, such as water turbidity/transparency, the distance between vegetation canopies and the water surface, and SAV coverage. Liew et al. proved that the spectral curve of SAV can change with the change in water turbidity and water depth [43]. They found that the typical NIR peak of vegetation spectra could not be detected at a water depth of 1 m with turbidity 0.5 NTU, and vegetation could not be detected at 0.5 m with high turbidity (50 NTU). Beget et al. found that the reflectance in NIR of the flooded vegetation decreased as its flooding level increased [21]. Variation in these factors might lead to the change of suitable ranges of the decision function. In order to explore the applicable conditions of the concave–convex decision function under different transparency, vegetation depth, and coverage, we integrated field hyperspectral data to four bands of GF-1 based on Equation (1). This convolution method was based on the band range of GF-1, and the NIR ranged from 0.77–0.89 μm. Various remote sensing sensors may have slightly different calculation results because of their different band ranges. Therefore, the applicability of this method to other sensors still needs further exploration.

As shown in Figure 10, at 0.3 m with 0.6 m transparency, although the NIR peak disappeared, the spectral curve of SAV was still concave in the third band. Thus, SAV could be judged by using the concave–convex decision function in this situation. However, this concave characteristic disappeared at 0.5 m. Overall, in a water body with a transparency of 0.6 m, the concave–convex decision function could be applied to 100% SAV coverage above 0.3 m. Below 0.3 m, both the water body and the SAV showed a convex shape in the third band, and they could be determined based on the included angle of the concave–convex decision function. In a 1.5-m transparency water body, the vegetation NIR peak still existed at 0.3 m due to SAV coverage of 100% and the high transparency of the water body, but it disappeared at 0.5 m. The range in which the SAV spectral curve was concave in the third band should be between 0.3 m and 0.5 m, but we did not capture the specific value because of the large interval setting of depth. As described in the results, SAV with 80% and 60% coverage in Figure 11 did not exhibit NIR high peak characteristics at 0.3 m. However, the SAV spectral curve was still concave in the third band at this depth. SAV could also be judged by using the concave–convex decision function. The concave shape of the SAV spectral curves in the third band both disappeared at 0.5 m. With 40% SAV coverage, NIR showed high reflectance above 0.1 m, but at 0.3 m, the concave shape of the SAV spectral curve in the third band disappeared. Therefore, when SAV coverage was less than 40%, even at 0.3 m, it is difficult to identify the SAV based only on the concave shape of the spectral curve. Overall, the concave–convex decision function could be applied to 80% and 60% SAV coverage above 0.3 m and 40% SAV coverage above 0.1 m under 1.5 m transparency.

5. Conclusions

It is almost impossible to accurately identify plants that grow underwater using conventional extraction methods for aquatic vegetation. The concave–convex decision function method proposed in this study could further accurately classify SAV from water bodies. When comparing the results of the concave–convex decision function method with the NDVI classification using the same data, the concave–convex decision function method clearly outperformed the NDVI classification. The decision function can be applied to waters with SAV coverage greater than 40% above 0.3 m and

Remote Sens. **2018**, *10*, 1279

SAV coverage 40% above 0.1 m under 1.5 m transparency. With 100% SAV coverage under 0.6 m transparency, the concave–convex decision function can be applied up to 0.3 m. Combining the concave–convex decision function flexibly in the classification method (e.g., using a decision tree) can achieve the accurate extraction of SAV, and provides new ideas for the accurate extraction of SAV in other regions.

Another outcome of interest from this research is the potential utility of the GF-1 in aquatic vegetation classification. When aquatic vegetation information from two-period GF-1 remote sensing images in July and August 2015 was classified using the decision tree method from Lake Ulansuhai, China, the overall accuracy was 92.17% and 91.79%, respectively. Four bands from GF-1 (a satellite with higher resolution than TM and HJ-1A/B, a shorter revisit period, and good continuity) had relatively good applicability for information extraction of aquatic vegetation. High-resolution GF-1 images combined with a new decision function were able to provide a simple and effective method to dynamically and accurately monitor aquatic vegetation, especially SAV, on a large regional scale, and could provide support for long-term ecosystem health monitoring.

Author Contributions: Q.C. conceived the idea of this study and wrote the paper. R.Y. and Y.H. directed the study. R.Y. and Q.C. designed the experiments. R.Y., Y.H., Q.C., W.Z., Q.Z., and X.B. carried out the field experiments. R.Y., Y.H., and L.W. revised the paper. R.Y. and Y.H. directed the major revision.

Funding: This research was funded by the National Natural Science Foundation of China (Grant Nos. 51469018, 61461034, 91547110 and 41701281); the National Key Research and Development Program of China (Grant No. 2016YFC0500508); and Ministry of Water Resources Special Funds for Scientific Research Projects of Public Welfare Industry (Grant No. 20150104).

Acknowledgments: We are indebted to two anonymous reviewers for their efforts and constructive comments. These comments and suggestions have greatly improved our work.

Conflicts of Interest: The authors declare no conflict of interest.

References

1. Jin, X. *Lake Environment in China*; China Ocean Press: Beijing, China, 1995.
2. Ma, R. *Remote Sensing of Lake Water Environment*; Science Press: Beijing, China, 2010.
3. Silva, T.S.F.; Costa, M.P.F.; Melack, J.M.; Novo, E.M.L.M. Remote sensing of aquatic vegetation: Theory and applications. *Environ. Monit. Assess.* **2008**, *140*, 131–145. [CrossRef] [PubMed]
4. Marshall, T.R.; Lee, P.F. Mapping aquatic macrophytes through digital image analysis of aerial photographs: An assessment. *J. Aquat. Plant Manag.* **1994**, *32*, 61–66.
5. Welch, R.; Remillard, M.M.; Slack, R.B. Remote sensing and geographic information system techniques for aquatic resource evaluation. *Photogramm. Eng. Remote Sens.* **1988**, *54*, 177–185.
6. Zhang, Y.; Liu, X.; Qin, B.; Shi, K.; Deng, J.; Zhou, Y. Aquatic vegetation in response to increased eutrophication and degraded light climate in eastern lake Taihu: Implications for lake ecological restoration. *Sci. Rep.* **2016**, *6*. [CrossRef] [PubMed]
7. Fusilli, L.; Collins, M.O.; Laneve, G.; Palombo, A.; Pignatti, S.; Santini, F. Assessment of the abnormal growth of floating macrophytes in Winam Gulf (Kenya) by using MODIS imagery time series. *Int. J. Appl. Earth Obs. Geoinf.* **2013**, *20*, 33–41. [CrossRef]
8. Ackleson, S.G.; Klemas, V. Remote sensing of submerged aquatic vegetation in lower Chesapeake Bay: A comparison of Landsat MSS to TM imagery. *Remote Sens. Environ.* **1987**, *22*, 235–248. [CrossRef]
9. Armstrong, R.A. Remote sensing of submerged vegetation canopies for biomass estimation. *Int. J. Remote Sens.* **1993**, *14*, 621–627. [CrossRef]
10. Luo, J.; Li, X.; Ma, R.; Li, F.; Duan, H.; Hu, W.; Qin, B.; Huang, W. Applying remote sensing techniques to monitoring seasonal and interannual changes of aquatic vegetation in Taihu Lake, China. *Ecol. Indic.* **2016**, *60*, 503–513. [CrossRef]
11. Xie, Y.; Sha, Z.; Yu, M. Remote sensing imagery in vegetation mapping: A review. *J. Plant Ecol.* **2008**, *1*, 9–23. [CrossRef]
12. Dogan, O.K.; Akyurek, Z.; Beklioglu, M. Identification and mapping of submerged plants in a shallow lake using quickbird satellite data. *J. Environ. Manag.* **2009**, *90*, 2138–2143. [CrossRef] [PubMed]

13. Pu, R.; Bell, S. Mapping seagrass coverage and spatial patterns with high spatial resolution IKONOS imagery. *Int. J. Appl. Earth Obs. Geoinf.* **2017**, *54*, 145–158. [CrossRef]

14. Whiteside, T.; Bartolo, R. Mapping aquatic vegetation in a tropical wetland using high spatial resolution multispectral satellite imagery. *Remote Sens.* **2015**, *7*, 11664–11694. [CrossRef]

15. Bolpagni, R.; Bresciani, M.; Laini, A.; Pinardi, M.; Matta, E.; Ampe, E.M.; Giardino, C.; Viaroli, P.; Bartoli, M. Remote sensing of phytoplankton-macrophyte coexistence in shallow hypereutrophic fluvial lakes. *Hydrobiologia* **2014**, *737*, 67–76. [CrossRef]

16. Wang, Y.; Traber, M.; Milstead, B.; Stevens, S. Terrestrial and submerged aquatic vegetation mapping in fire island national seashore using high spatial resolution remote sensing data. *Mar. Geod.* **2007**, *30*, 77–95. [CrossRef]

17. Lin, C.; Gong, Z.; Zhao, W. The extraction of wetland hydrophytes types based on medium resolution TM data. *Acta Ecol. Sin.* **2010**, *30*, 6460–6469.

18. Zhao, D.; Lv, M.; Jiang, H.; Cai, Y.; Xu, D.; An, S. Spatio-temporal variability of aquatic vegetation in Taihu Lake over the past 30 years. *PLoS ONE* **2013**, *8*, 10454–10461. [CrossRef] [PubMed]

19. Davranche, A.; Lefebvre, G.; Poulin, B. Wetland monitoring using classification trees and spot-5 seasonal time series. *Remote Sens. Environ.* **2010**, *114*, 552–562. [CrossRef]

20. Ma, R.; Duan, H.; Gu, X.; Zhang, S. Detecting aquatic vegetation changes in Taihu Lake, China using multi-temporal satellite imagery. *Sensors* **2008**, *8*, 3988–4005. [CrossRef] [PubMed]

21. Beget, M.E.; Bella, C.M.D. Flooding: The effect of water depth on the spectral response of grass canopies. *J. Hydrol.* **2007**, *335*, 285–294. [CrossRef]

22. Zhang, S.; Duan, H.; Gu, X. Remote sensing information extraction of hydrophytes based on the retrieval of water transparency in lake Taihu, China. *J. Lake Sci.* **2008**, *20*, 184–190. [CrossRef]

23. Li, F.; Xiao, B. Aquatic vegetation mapping based on remote sensing imagery: An application to Honghu Lake. In Proceedings of the International Conference on Remote Sensing, Environment and Transportation Engineering, Nanjing, China, 24–26 June 2011; pp. 4832–4836.

24. Visser, F.; Wallis, C.; Sinnott, A.M. Optical remote sensing of submerged aquatic vegetation: Opportunities for shallow clearwater streams. *Limnol. Ecol. Manag. Inland Waters* **2013**, *43*, 388–398. [CrossRef]

25. Alberotanza, L. Hyperspectral aerial images. A valuable tool for submerged vegetation recognition in the Orbetello Lagoons, Italy. *Int. J. Remote Sens.* **1999**, *20*, 523–533. [CrossRef]

26. Li, J.; Wu, D.; Wu, Y.; Liu, H.; Shen, Q.; Zhang, H. Identification of algae-bloom and aquatic macrophytes in Lake Taihu from in-situ measured spectra data. *J. Lake Sci.* **2009**, *21*, 215–222.

27. Cho, H.J.; Lu, D. A water-depth correction algorithm for submerged vegetation spectra. *Remote Sens. Lett.* **2010**, *1*, 29–35. [CrossRef]

28. Cho, H.J.; Mishra, D.; Wood, J. Remote sensing of submerged aquatic vegetation. In *Remote Sensing—Applications*; Escalante, B., Ed.; InTech: Rijeka, Croatia, 2012.

29. Wang, L.; Yang, R.; Tian, Q.; Yang, Y.; Zhou, Y.; Sun, Y.; Mi, X. Comparative analysis of GF-1 WFV, ZY-3 MUX, and HJ-1 CCD sensor data for grassland monitoring applications. *Remote Sens.* **2015**, *7*, 2089–2108. [CrossRef]

30. Xia, C.; Zhang, Y.; Wang, W. A relief-based forest cover change extraction using GF-1 images. In Proceedings of the IGARSS 2014—2014 IEEE International Geoscience and Remote Sensing Symposium, Quebec, QC, Canada, 13–18 July 2014.

31. Yang, Y.; Zhan, Y.; Tian, Q.; Gu, X.; Yu, T.; Wang, L. Crop classification based on GF-1/WFV NDVI time series. *Trans. Chin. Soc. Agric. Eng.* **2015**, *31*, 155–161.

32. He, L.; Xi, B.; Lei, H. *Research on Integrated Treatment and Management Planning of Lake Ulansuhai*; China Environmental Science Press: Beijing, China, 2013.

33. Zheng, W.; Han, X.; Liu, C. Satellite remote sensing data monitoring "Huang Tai" algae bloom in lake Ulansuhai, inner Mongolia. *J. Lake Sci.* **2010**, *22*, 321–326.

34. Jia, K.; Liang, S.; Gu, X.; Baret, F.; Wei, X.; Wang, X.; Yao, Y.; Yang, L.; Li, Y. Fractional vegetation cover estimation algorithm for chinese GF-1 wide field view data. *Remote Sens. Environ.* **2016**, *177*, 184–191. [CrossRef]

35. Wu, Z.; Yang, F.; Zhang, Y.; Wu, Y.; Yu, W. Quality evaluation of GF-1 and SPOT-7 multi-spectral image based on land surface parameter validation. *J. Image Graph.* **2016**, *21*, 1551–1561.

36. Zhao, Y. *Principles and Methods of Analysis of Remote Sensing Applications*; Science Press: Beijing, China, 2003.

Remote Sens. **2018**, *10*, 1279

37. Han, L.; Rundquist, D.C. The spectral responses of *Ceratophyllum demersum* at varying depths in an experimental tank. *Int. J. Remote Sens.* **2003**, *24*, 859–864. [CrossRef]
38. Cho, H.J.; Kirui, P.; Natarajan, H. Test of multi-spectral vegetation index for floating and canopy-forming submerged vegetation. *Int. J. Environ. Res. Public Health* **2008**, *5*, 477. [CrossRef] [PubMed]
39. Cho, H.J. Depth-variant spectral characteristics of submersed aquatic vegetation detected by Landsat 7 ETM+. *Int. J. Remote Sens.* **2007**, *28*, 1455–1467. [CrossRef]
40. Cohen, J. A coefficient of agreement for nominal scales. *Educ. Psychol. Meas.* **1960**, *20*, 37–46. [CrossRef]
41. Cohen, J. Weighted kappa: Nominal scale agreement with provision for scaled disagreement or partial credit. *Psychol. Bull.* **1968**, *70*, 213–220. [CrossRef] [PubMed]
42. Jensen, J.R. *Introductory Digital Image Processing a Remote Sensing Perspective*; Science Press: Beijing, China, 2007.
43. Liew, S.C.; Chang, C.W. Detecting submerged aquatic vegetation with 8-band worldview-2 satellite images. In Proceedings of the Geoscience and Remote Sensing Symposium, Munich, Germany, 22–27 July 2012; pp. 2560–2562.

remote sensing

MDPI

Article

Using 1st Derivative Reflectance Signatures within a Remote Sensing Framework to Identify Macroalgae in Marine Environments

Ben Mcilwaine, Monica Rivas Casado * and Paul Leinster

School of Water, Energy and Environment, Cranfield University, College Road, Cranfield MK43 0AL, UK;
ben.mcilwaine@cranfield.ac.uk (B.M.); paul.leinster@cranfield.ac.uk (P.L.)
* Correspondence: m.rivas-casado@cranfield.ac.uk; Tel.: +44-012-3475-0111

Received: 28 January 2019; Accepted: 20 March 2019; Published: 23 March 2019

Abstract: Macroalgae blooms (MABs) are a global natural hazard that are likely to increase in occurrence with climate change and increased agricultural runoff. MABs can cause major issues for indigenous species, fish farms, nuclear power stations, and tourism activities. This project focuses on the impacts of MABs on the operations of a British nuclear power station. However, the outputs and findings are also of relevance to other coastal operators with similar problems. Through the provision of an early-warning detection system for MABs, it should be possible to minimize the damaging effects and possibly avoid them altogether. Current methods based on satellite imagery cannot be used to detect low-density mobile vegetation at various water depths. This work is the first step towards providing a system that can warn a coastal operator 6–8 h prior to a marine ingress event. A fundamental component of such a warning system is the spectral reflectance properties of the problematic macroalgae species. This is necessary to optimize the detection capability for the problematic macroalgae in the marine environment. We measured the reflectance signatures of eight species of macroalgae that we sampled in the vicinity of the power station. Only wavelengths below 900 nm (700 nm for similarity percentage (SIMPER)) were analyzed, building on current methodologies. We then derived 1st derivative spectra of these eight sampled species. A multifaceted univariate and multivariate approach was used to visualize the spectral reflectance, and an analysis of similarities (ANOSIM) provided a species-level discrimination rate of 85% for all possible pairwise comparisons. A SIMPER analysis was used to detect wavebands that consistently contributed to the simultaneous discrimination of all eight sampled macroalgae species to both a group level (535–570 nm), and to a species level (570–590 nm). Sampling locations were confirmed using a fixed-wing unmanned aerial vehicle (UAV), with the collected imagery being used to produce a single orthographic image via standard photogrammetric processes. The waveband found to contribute consistently to group-level discrimination has previously been found to be associated with photosynthetic pigmentation, whereas the species-level discriminatory waveband did not share this association. This suggests that the photosynthetic pigments were not spectrally diverse enough to successfully distinguish all eight species. We suggest that future work should investigate a Charge-Coupled Device (CCD)-based sensor using the wavebands highlighted above. This should facilitate the development of a regional-scale early-warning MAB detection system using UAVs, and help inform optimum sensor filter selection.

Keywords: macroalgae; reflectance; 1st derivative; species discrimination; unmanned aerial vehicle; nuclear power station

1. Introduction

Algal blooms are the cause of large-scale damage and disruption to coastal operators [1], including power generation plants whose water intakes can get blocked, or mechanically damaged [2]. In France, 3.6 million francs were spent on the removal of 90,000 m^3 of microalgae "green tides" in 1992, while in Lee County (USA) a total of $260,500 was spent in 2003/2004 to address problems caused by Rhodophyta blooms, and in Australia $160,000 are spent every year removing around 13,000 m^3 of macroalgae [1]. Microalgal blooms are well known for their propensity to generate 'red tides' as well as their strong links to harmful algal blooms (HABs) [3–5]. These microalgae blooms are generated by the discharge of excess nutrients into water bodies [6–8]. The toxins produced by these algae can kill marine mammals, fish and other vertebrates via food chain biomagnification of toxins [4,9]. Microalgae blooms cause biological damage to shellfish farms, induce localized ecosystem disruption and foul desalination plants [10–12]. In addition, macroalgae blooms (MABs) are known to cause significant environmental and economic damage, especially if their extent leads to aquatic hypoxic conditions due to a lack of dissolved oxygen [13], resulting in catastrophic ecosystem collapse. MABs form through large-scale detachment from their growth location resulting in their suspension within the water column [8,14]. This transition from being sessile, to being mobile, plays a key role in the generation of damaging blooms. MABs also have an impact on indigenous species, nuclear power stations and fish farms [1,14], particularly when amassing to sizes over 0.50 km^2 [14]. Assuming a macroalgae mass of 1 kg m^{-2}, this would suggest a bloom mass of around 560 tons. These macroalgae aggregations have the potential to disrupt impacted industries predominantly via non-biotoxin mechanisms.

The characteristics of microalgal blooms have been well researched. However, the causes and effects of MABs are less well understood [3,15]. Despite a heightened pressure on affected industries via social, economic, and underlying ecological trends [1,14], MAB research is still currently minimal [16]. If the issues caused by MABs are to be addressed, then appropriate monitoring and surveillance methodologies are required. Remote sensing clearly has an important role to play in such methodologies and would require a comprehensive understanding of the spectral characteristics of species that can detach from substrates and form MABs [5]. For the remote sensing warning system to be effective, 6–8 h alert of an impending ingress event is required (EDF, personal comment).

A considerable amount of work has already been undertaken using airborne and space borne techniques to detect high density, surface or shallow water (less than 13 m [17]) sessile submerged aquatic vegetation (SAV) [14,17–21]. MABs have in fact been successfully detected on the ocean surface with the use of satellite-based SAR and spectral radiometers [14,22]. The authors identified limitations associated with these techniques; SAR was not able to penetrate the ocean surface and spectral radiometers did not function if cloud was present. In addition, the resolution of such systems would also too low to detect low-density MABs that can still cause damage [21]. The time taken to collect and process the data is also a factor due to inherent satellite data latency [23,24]. Satellite data is therefore not considered at present to be a practical means of providing warnings within the 6 to 8 h time frame required by coastal operators. MABs have also been tracked using a range of morphological, physiological, and molecular techniques [8]. However, these methods do not allow surveys to be carried out over large areas, frequent monitoring to be undertaken or near real time analysis. UAVs can enable rapid deployment within a specified location with ready data access, which should provide the means to warn of a potentially damaging event with enough time to act. This is critical within the context of a regional early-warning system.

The ability of coastal industries to introduce appropriate mitigation measures to minimize the impacts of recurrent MABs requires appropriate surveillance methodologies including identifying bloom generation and detachment [8]. By identifying the spectral reflectance signature of the problem species, and focusing on the characteristic spectral reflectance bands that can also penetrate water, we should be able to gain more information about bloom composition. This can then be used to develop an early-warning system that will enable coastal operators to minimize damage to their process equipment.

The characterization of vegetation spectral signatures has been successfully used to differentiate between oceanic surface conditions as a predictor for microalgae bloom presence [25], for large-scale monitoring and detection of SAV to aid ecological engineering efforts [26], and to measure temporal changes over long time periods. However, little is yet known about the spectral reflectance signatures of MABs. There has been substantial research into the detection of terrestrial vegetation [27–30] but there are only a few papers on the remote sensing of low-density, varying-depth, mobile macroalgae using their spectral reflectance signatures [31–34]. Seagrasses have been thoroughly researched by [18–20,35] and successfully differentiated into three different species by [32], however these are taxonomically plants and not macroalgae. This study used wavelengths between 530–580 nm with "additional discrimination" provided from 520–530 nm and 580–600 nm, in addition to an absorption trough at 686–700 nm (using red pigments). They found that wavelengths between 550–560 nm and 700–710 nm were most sensitive to chlorophyll detection; it is likely, due to the similarities between seaweeds, fucoids, and seagrasses that similar wavebands will be useful for MABs, as photosynthetic species have similar pigment structures [32] and in turn spectral reflectance characteristics. Species can be differentiated through these characteristic photosynthetic pigment reflectance signals. The relative absorption characteristics at different wavelengths can vary greatly between species with age, seasonal cycles, growth stage and genetic variation all affecting the absorption profile [32]. However, it has been found that seagrass species were able to be identified in the presence of other species even if fouled, irrespective of spatial and temporal variability [32]. It may therefore be possible to use the spectral reflectance of vegetation to develop a remote sensing technique for the reliable detection of mobile MAB presence. We aim to identify areas of maximum spectral separation. Once these are determined, they can be used to inform sensor selection and filter optimization on regional-scale remote sensing platforms. The practicalities of the chosen sensor type can then be explored in further detail as exemplified by [36]. The work presented here has the potential to contribute to the development of more robust monitoring methods and programs for the early detection of seaweed ingress.

The aim of this study is to identify the spectral reflectance signatures of the macroalgae that have been responsible for adverse impacts on coastal power generation plants. This will be achieved through the following objectives:

1. Ascertain the reflectance signature of species within the functional macroalgae groups found during sampling at the site of interest.
2. Quantify the differences in spectral reflectance profiles between sampled macroalgae groups.
3. Identify and discriminate between sampled macroalgae groups based on the results from (1) and (2).

2. Materials and Methods

2.1. Site Selection

The study site is located near Torness nuclear power station (East Lothian, UK). The power station is one of the UK's second-generation nuclear reactors, powered by two advanced gas-cooled reactors, and has four drum screens within a single cooling water intake. The location was chosen due to the site's susceptibility to disruption caused by ingress of large masses of macroalgae. This has resulted in the energy company suffering significant revenue losses each year. Each emergency shut down costs the company around $2 million per day [37].

2.2. Data Collection

A total of 15 kg of macroalgae were collected for analysis. Field sampling was conducted during the last week of June 2018 at a beach in East Lothian, UK under blue sky conditions with a maximum temperature of 32 °C. Prior to sample collection, the study site was explored on foot to indicate areas of high macroalgae density. However, sampling efforts were restricted by incoming tides and restricted access points due to proximity of the nuclear site. The areas of high biomass density were confirmed with UAV flights (Figure 1) in case any areas may have been missed. Samples were collected via stratified sampling based on biomass dominance. Collected samples were stored in plastic bags, maintained cool in portable refrigerators and transported to a refrigeration unit within eight hours of collection.

Optimum macroalgae sampling locations were determined using a fixed-wing Intel Sirius Pro Unmanned Aerial Vehicle (UAV). Using a Sony Alpha 6300 camera, 3962 RGB aerial images were collected over the course of three flight missions. Each mission used the Intel advanced flight planning software: MAVinci desktop (MAVinci, St. Leon-Rot, Germany). Each mission had a pre-determined flight path that was optimized for maximum spatial coverage of the area surrounding the nuclear power station while maintaining enough resolution to identify seaweed coverage. Each mission was flown at a height of 100 m which resulted in a ground sampling distance (GSD) of 2 cm for all conducted flights. The camera used a 23.5 × 15.6 mm complementary metal-oxide-semiconductor (CMOS) sensor, with a maximum resolution of 24 MP and an ISO range of 100–25,600. Of the 3962 collected images, 2788 were selected to be used for photogrammetric analysis. For the generation of the orthoimage (Figure 1), Photoscan Pro version 1.1.6 (Agisoft LLC, St. Petersberg, Russia) was used to stitch the images together. The resultant orthoimage was assessed to finalize sampling locations that had the optimum probability of high macroalgae densities. As a result, of the sampling protocol, the research was focused on the macroalgae groups that are most likely to significantly contribute to disruption at the Torness nuclear power station (Figure 1).

This paper builds on the methods employed by [32] who found that the spatial and temporal variability of each species did not affect species discrimination. Based on these findings, both spatial and temporal variation were considered but not included within the sampling procedure.

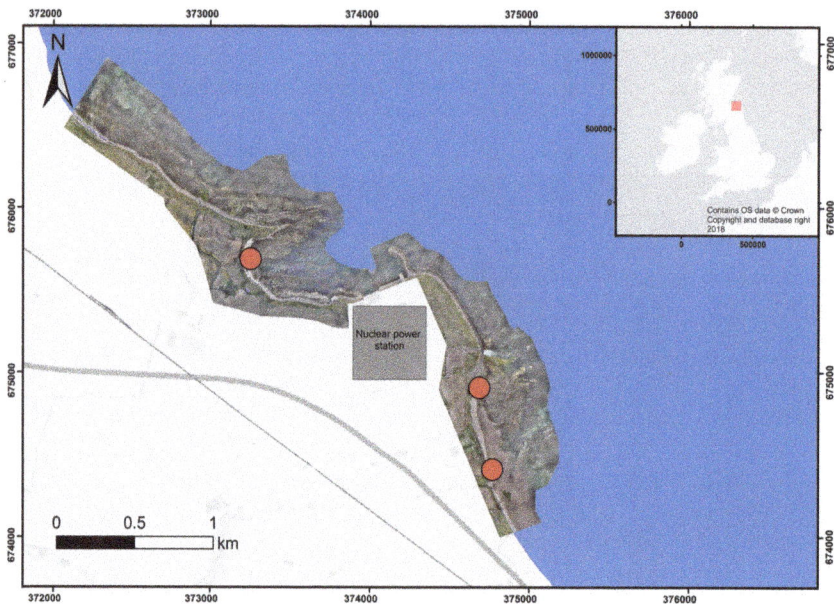

Figure 1. Hybrid map of the study site with orthoimage of the coastline embedded. The nuclear power station is Torness, Scotland, UK (EDF Energy). Red markers show the location of the sampling sites. Coordinates used from the British National Grid system. *Contains OS data © Crown copyright and database.*

2.3. Laboratory Sampling of Spectral Reflectance

Spectral readings were taken over five consecutive days using an Analytical Spectral Devices (ASD) FieldSpec 4 HI-Res spectroradiometer that records radiance with 2151 channels, a spectral range of 350–2500 nm, and with resolutions as follows: visible and near-infrared (VNIR) 3 nm (at 700 nm), and short-wavelength infrared (SWIR) 8 nm (at 1400/2100 nm). The 'FieldSpec 4' was calibrated via a Spectralon SRM-99 [38], being the most optically appropriate reference panel for the spectral range of the spectroradiometer used. Readings were taken in ex-situ laboratory conditions using an ASD 'contact probe' thereby eliminating the influence of background light sources while in contact with the desired target; the probe provides its own regulated and controlled light source. The spectroradiometer was run for one hour prior to taking readings in accordance with [39] who recommended this procedure to obtain reliable and comparable results.

Spectroradiometric measurements were taken with the measurement wand in contact with the sample, and seaweed samples were not dehydrated prior to recording of spectra. There are benefits to dehydrating vegetation prior to taking spectral reflectance readings [40]. However, for marine species this is not advised. Dehydration of seaweed samples would provide spectral reflectance information that would not be relevant in their full marine habitat. Keeping the sample moist while taking reflectance readings should therefore be common practice when dealing with SAV [32]. Analysis was only focused on wavelengths between 400–900 nm, building on the methods used by [20,21,32,41] who focused on lower wavelengths due to "high absorption of light in the water column" [41]. Although higher wavelengths can be used to successfully detect macroalgae found in shallow waters, a regional-scale early-warning detection system must be able to detect seaweeds found in deeper waters as well [42].

The total number of samples per species were not equal due to the relative presence of species at the sampling locations. The number of spectral readings per functional group were: kelp, 1522, fucoid, 1130, other, 381. Kelp samples were cut into as many 30 mm pieces as possible (just bigger than the contact probe head) for ease of handling and to maximize the number of sample readings; this ensured that consideration was given to the intra-specific color variation between samples taken from the same species. Each sample was subject to a single reading taken on top of non-reflective black background as per [41]. Due to the morphological differences compared to kelp, fucoid species were not cut. Fucoids were laid flat and readings were taken at every intersecting point on a grid consisting of 40 × 40 mm squares. This process ensured no readings overlapped and independence of data was maintained.

The FieldSpec 4 provided an output of reflectance at each wavelength per spectral reading. The FieldSpec 4's "spectral averaging" setting can automatically average multiple readings to provide a single output. To achieve highly smoothed spectral outputs, it is suggested within the FieldSpec 4 field guide [39] to select between 15–25 spectra to be averaged per output. For added statistical robustness, we decided to use 50 averaged signals. In combination with a controlled light source, this ensured that a naturally smooth spectral profile was produced. Outputs per cut piece of seaweed were then processed to provide insight into the overall spectral reflectance signal.

2.4. Data Analysis

IndicoPro Ver. 6.4 (Malvern Panalytical, Malvern, UK) [43] was used alongside the ASD FieldSpec 4 to extract the raw spectra with the software ViewSpecPro Ver. 6.2.0 (Malvern Panalytical, Malvern, UK) [43] being used for the post-processing of the collected spectra. Post-processing steps included: visual overlaying of spectra, averaging of spectra for initial visualization, and data extraction to ASCII file format (Figure 2). The sampled spectral reflectance values were converted and exported as 1st derivative spectra to reduce the effect of amplitude variation between sample readings and emphasize areas of spectral change [44].

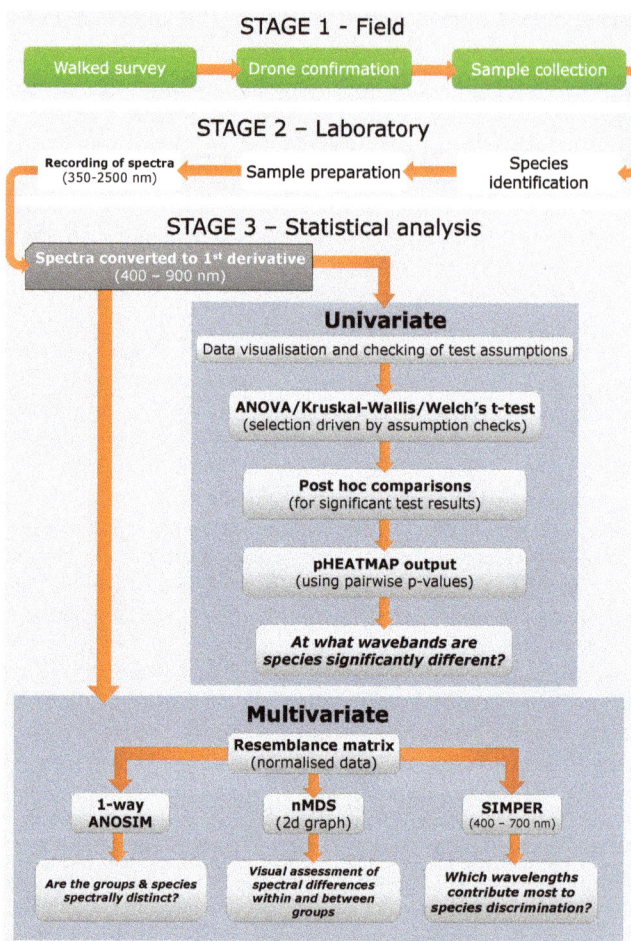

Figure 2. Overall research flow chart summarizing the methodological processes of the field work, laboratory, and statistical analysis procedures.

2.4.1. Inter-Specific Spectral Differences

A one-way analysis of variance (ANOVA) was conducted at each wavelength from 400–900 nm, to provide evidence of where statistically significant differences in reflectance of macroalgae species occurs. Prior to running each one-way ANOVA per wavelength, the data were checked for normality and homoscedasticity. For non-normally distributed data, a Fligner-Killeen test [45] for homogeneity of variances was completed followed by a Kruskal-Wallis H test [46]. Post-hoc comparison tests were then conducted (if significant differences were found) with a holm adjustment to account for additional risk of type 1 errors. If the data were found to be normally distributed, then a Bartlett's test [47] was conducted. Data found to lack homoscedasticity were subjected to a Welch's *t*-test [48], again with post-hoc tests completed to find which specific combinations were significantly different. If data were found to be normally distributed while retaining homoscedasticity, then the one-way ANOVA was completed with a post-hoc Tukey test [49] for unequal sample sizes. The package "pheatmap" [50] was used to create a graphical representation of the significance of every pairwise comparison at each wavelength. All univariate tests were conducted using the statistical software R 3.4.3 [51] with the following packages: "vegan" [52], "ggplot2" [53] and "reshape2" [54].

To conduct multivariate analysis, all data were normalized and a resemblance matrix produced using Euclidean distances due to the presence of negative data values from the 1st derivative data [55]. The matrix was then used to produce a 2-dimensional output (Figure 3) of the multidimensional data via non-metric multidimensional scaling (nMDS). The nMDS was conducted to visually assess the differences between spectral signatures both within, and between, the broader macroalgae groups as well as at a species level.

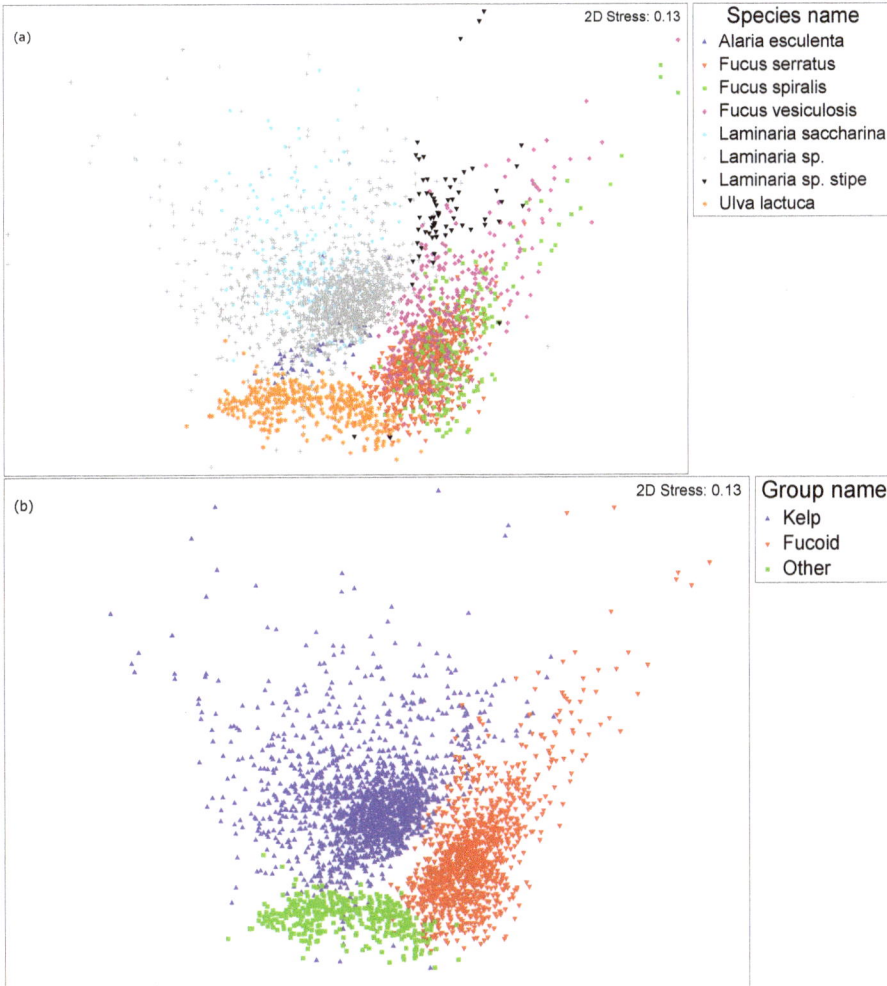

Figure 3. Non-metric multidimensional scaling (nMDS) visualizing the variance in spectral reflectance expression: (**a**) Between species. (**b**) Between groups.

2.4.2. Formal Testing of Spectral Differences between Groups (and Species) with ANOSIM

Using the resemblance matrix, a one-way analysis of similarity (ANOSIM) [56] was conducted to determine whether there were significant differences present between the broader groups of macroalgae sampled, as well as all possible species comparisons. The ANOSIM test uses ranked dissimilarity values of the 1st derivative data within the resemblance matrix. As an ANOSIM is a distribution free, non-parametric test with no assumptions of homogeneity of variances or normality of data, no testing of these assumptions was completed.

The critical output of an ANOSIM is the R statistic. Values for the R statistic theoretically range between −1 and 1; however, in reality they range from 0 to 1. This is because negative R statistic values would suggest that differences within groups are greater than between groups. Any positive R statistic values suggest that there is dissimilarity between groups. A value of R = 0 suggests no dissimilarity between groups, and R = 1 suggests complete dissimilarity. The ANOSIM analysis calculates a scenario in which there are no differences between tested groups, and what the R value output is for each of these 999 permutations (called R′). If the true R value is larger than any of these 999 R′ values, it can then be treated as a rare event (minimum 1 in 1000 chance). This therefore allows the rejection of the null hypothesis, that there are no differences between groups, to be rejected at $p < 0.001$. The true R value can be treated as a measure of absolute difference between groups, providing an indication of the magnitude of dissimilarity for a specific comparison. When used in combination with nMDS, a more informed analysis of group dissimilarity can occur due to the formal significance of the ANOSIM complementing the visualization of the nMDS [57].

2.4.3. Wavelength Analysis to Find the Best Discriminating Wavelengths (SIMPER)

Similarity percentage (SIMPER) analysis [32,56] was undertaken to identify which wavelengths were the highest contributors to any significant spectral variation between individual species. The SIMPER analysis evaluates the contribution of each wavelength to the observed dissimilarity between species via reflectance. The resulting output allows us to identify which wavelengths are most critical in any observed patterns of differentiation. If a specific wavelength is consistently providing high levels of within species similarity—a metric for being characteristic of the species—in addition to between group dissimilarity, then that wavelength will be able to be used for reliable species discrimination [56]. Only wavelengths below 700 nm were investigated because of the dominance of the near-infrared (NIR) wavebands within the SIMPER analysis. It is these lower wavelengths that have greater water penetration capability [42]. The wavelengths between 700–900 nm dominated and prevented the detection of the lower discriminatory wavelengths. This dominant reflectance in NIR bands [58] provides an unhelpful detection bias towards surface and shallow marine habitats. In the context of an early-warning detection system for potentially dangerous macroalgae blooms, it is not suitable to only have the capability to detect the upper sections of the water column. Within the output of the SIMPER analysis, the wavelengths that contribute most to differentiation are found to have the highest "Sum % contribution" values for a given wavelength; a metric for the influence a specific wavelength is having on the discrimination of all species (or group) comparisons. All multivariate analysis was completed using PRIMER v7 [55].

3. Results

3.1. Laboratory Sampling of Spectral Reflectance

Three groups of SAV were sampled and their spectral reflectance properties analyzed. A total of 3033 readings were obtained, with the number of readings taken for each species shown in Table 1. The species composition of the three groups are shown below (Table 1). Species were identified using the Environment Agency seaweed reference manual [59]. Although not a true taxonomic species, for the ease of discussion and analysis, the samples of *Laminaria* sp. stipe are treated and referred to as a species. A plot of the mean with ± 1SD for each species' raw and 1st derivative spectra can be found in Appendix A.

Table 1. Groups of investigated species causing similar functional damage, the number of readings (n) are displayed in brackets next to each corresponding species.

Kelp	Fucoids	Other
Laminaria saccharina ($n = 230$)	*Fucus vesiculosus* ($n = 290$)	*Ulva lactuca* ($n = 381$)
Laminaria sp. ($n = 1177$)	*Fucus serratus* ($n = 612$)	
Laminaria sp. stipe ($n = 68$)	*Fucus spiralis* ($n = 228$)	
Alaria esculenta ($n = 47$)		

3.2. Data Analysis

3.2.1. Inter-Specific Spectral Differences

Pairwise tests were completed to investigate whether the sampled species (Table 1) were spectrally distinct when compared to all other possible combinations of species. With eight species sampled, 28 unique comparisons were available for testing. Prior to spectral sampling, it was observed that there were clear spectral differences in visible appearance between the three groups; however, these differences were not as noticeable within each group. While full spectra (350–2500 nm) were collected (Figure 2), only wavelengths from 400–900 nm (Figure 4) were analyzed due to the lack of practical application of the higher wavelengths; useable water penetration capability being a key requirement for the remote sensing of MABs. There were no broad wavebands (>30 nm) that had high levels of significance for all 28 pairwise comparisons (Figure 5). However, there were many narrow bands (<10 nm) that did exhibit high significance. These narrow wavebands have the potential to be used for species discrimination.

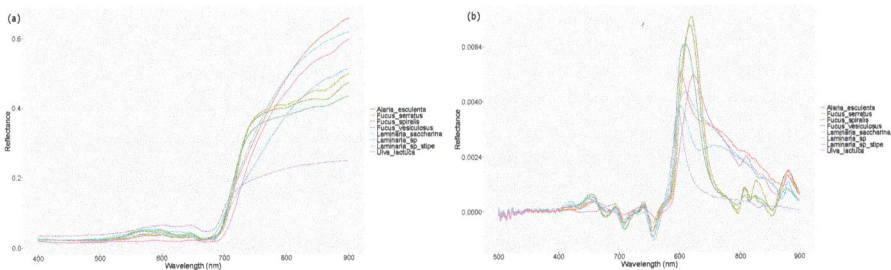

Figure 4. Averaged spectral reflectance per species: (**a**) Raw spectral reflectance. (**b**) 1st Derivative spectral reflectance.

Both *Fucus serratus* x *F. spiralis* and *F. vesiculosus* x *F. spiralis* comparisons (Figure 5) have poor discrimination at lower wavelengths but a highly significant band within the 500–600 nm range. This shows that even taxonomically and morphologically similar species can be differentiated with targeted wave band selection. There are some practically useful yet narrow bands that can simultaneously differentiate all 28 species comparisons. Conversely, there are areas of the spectrum that are clearly not appropriate for species-level spectral differentiation; 755–775 nm is a very poor area for comparing all three fucoid species to *Laminaria* sp. stipe and below 550 nm is also particularly poor for two of these comparisons (Figure 5). The most distinct combination is *L.* sp. and *F. serratus* (closely followed by *Ulva lactuca* x *L.* sp., *U. lactuca* x *F. vesiculosus* & *L. saccharina* x *F. serratus*) with strong levels of significance across the majority of the 400–900 nm spectrum. The three *L.* sp. stipe x fucoid comparisons have multiple broad areas of low significance within the 500–600 nm area, as well as the wavebands surrounding 775 nm. All comparisons for *U. lactuca* are highly significant across the spectrum suggesting that this was the most spectrally distinct species sampled. Wavelengths between 550–750 nm show the greatest range of significance for most comparisons, with 570–590 nm showing strong significance for all 28 comparisons.

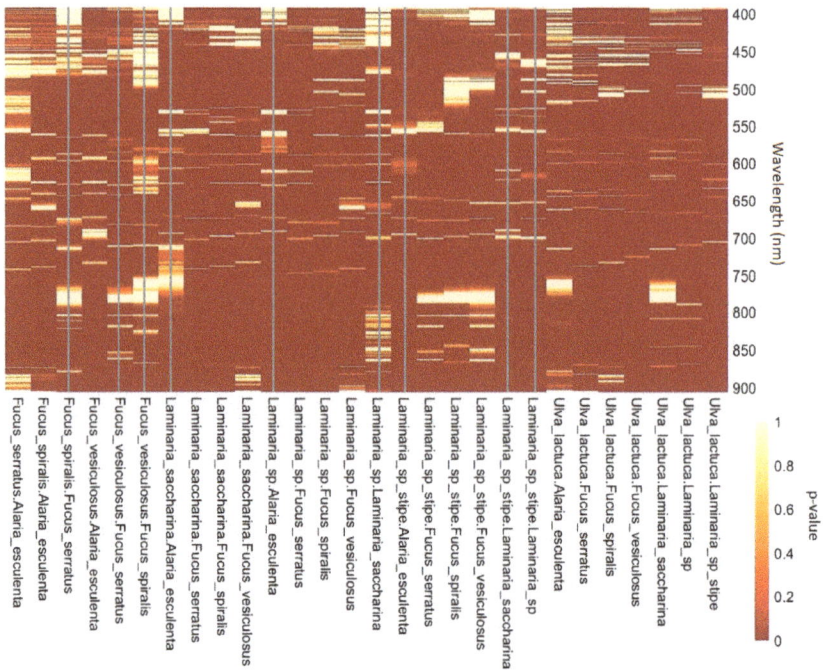

Figure 5. Heatmap of pairwise comparison *p*-values, intra-group comparisons represented by vertical bars. Legend depicting *p*-values with respect to heat intensity.

Multivariate analysis (Figure 3a) revealed strong spectral overlapping between all three fucoid species. There was also no clear distinction between the four kelp species. *U. lactuca* was clearly distinguishable from the other species, which supports the findings of the pairwise heat map (Figure 5). Within the four kelp species, there is complete spectral overlap present between *L. saccharina* and *L. sp.* (Figure 3a) indicating extreme levels of spectral similarity. The least sampled species *Alaria esculenta* (*n* = 47) shows a distinct cluster between the main groupings of *L. sp.* and *U. lactuca*. The *L.* sp. stipe readings display slight dissimilarity when compared with most other kelp species readings, with a marginal overlap with *F. vesiculosis* as well. However, it should be stressed that this is only due to some extreme values of *F. vesiculosus*.

Figure 3 enables some fundamental observations to be made. The two most spectrally distinct species are *L.* sp. stipe and *U. lactuca*, and the three fucoid species are spectrally similar while at the same time being distinct from all kelp species. *L. saccharina*'s spectral reflectance cannot be distinguished from the kelp species *L.* sp., while *A. esculenta* shares spectral similarity with the other kelp species but has a detectable level of spectral uniqueness. An nMDS stress value of 0.13 indicates that an accurate and reliable two-dimensional plot (Figure 3) [56] is being produced through the scaling of the multidimensional data set. The nMDS conducted supports the results shown by the pairwise heatmap (Figure 5) that there are significant differences between many of the inter-group species comparisons.

3.2.2. Formal Testing of Spectral Differences between Groups (and Species) with ANOSIM

The ANOSIM group analysis strongly rejects the null hypothesis that there are no spectral reflectance differences between the three sampled macroalgae groups (global R = 0.549, $p < 0.001$). An R value of this magnitude suggests that there are significant and distinctive differences in the spectral expression of all three groups (R = 0 meaning no differences, R = 1 meaning that all dissimilarity in spectral reflectance between groups is larger than any dissimilarity expressed within each group). The greatest difference in group spectral reflectance (Table 2) is between fucoid, and other (R = 0.712). This is followed by the kelp x fucoid comparison (R = 0.539), and lastly kelp x other (R = 0.455). These results support the findings of the nMDS plot (Figure 3b).

Table 2. ANOSIM group pairwise R values. All *p*-values < 0.001. All permutation values = 999.

	Kelp	**Fucoid**	**Other**
Kelp	-	-	-
Fucoid	0.539	-	-
Other	0.455	0.712	-

The ANOSIM analysis calculates 999 permutations (R′ values) for a scenario where there are no differences in group spectral reflectance, and then plots it against the true global R value. Due to stochastic variance between permutations, there are R′ values that vary around R = 0; however, none of them exceeded R′ = 0.025. Due to the true global R value (R = 0.549) being larger than any of the 999 R′ values, we can reject the null hypothesis with a certainty of *p* < 0.001.

Due to significant pairwise differences being found between sample groups, ANOSIM analysis was also run to a species level to investigate where the exact spectral differences were occurring. The ANOSIM species analysis strongly rejects the null hypothesis that there are no spectral reflectance differences between the eight sampled species (global R = 0.544, *p* < 0.001). Table 3 provides a deeper insight into the differences between spectral reflectance of each of the eight species.

Despite all pairwise comparisons (Table 3) with R values larger than the permutation maximum (R′ = 0.025), some comparisons still had large amounts of spectral similarity, indicated by their lower R values (highlighted in green). All intra-group species comparisons had R values of less than 0.49 which supports the findings of Figure 3a that there are observable spectral similarities within many of the intra-group species comparisons, especially between the fucoid species. *L.* sp. stipe maintains the most consistently strong levels of dissimilarity across all possible comparisons, followed closely by *U. lactuca*. Pairwise comparisons that exhibited notable spectral dissimilarity are highlighted in amber (Table 3).

Table 3. ANOSIM group pairwise R values. All *p*-values < 0.001. All permutation values = 999. Red cells highlight R > 0.66, amber cells highlight 0.67 > R > 0.33, green cells highlight 0.34 > R. All *p*-values ≤ 0.001 unless otherwise stated in brackets.

	Alaria esculenta	*Fucus serratus*	*Fucus spiralis*	*Fucus vesiculosus*	*Laminaria saccharina*	*Laminaria* sp.	*Laminaria* sp. stipe	*Ulva lactuca*
Alaria esculenta	-	-	-	-	-	-	-	-
Fucus serratus	0.885	-	-	-	-	-	-	-
Fucus spiralis	0.750	0.331	-	-	-	-	-	-
Fucus vesiculosus	0.631	0.457	0.140	-	-	-	-	-
Laminaria saccharina	0.451	0.930	0.824	0.707	-	-	-	-
Laminaria sp.	0.0617 (0.092)	0.583	0.610	0.492	0.159	-	-	-
Laminaria sp. stipe	0.795	0.969	0.786	0.559	0.872	0.685	-	-
Ulva lactuca	0.490	0.848	0.830	0.805	0.797	0.533	0.939	-

The cells highlighted in red exhibit comparisons that have exceptionally high levels of species differentiation with values of R > 0.66; *L.* sp. stipe comparisons are of particular note showing only one mid-range R value (with *F. vesiculosus*) which is also represented in Figure 3a with a slight overlap in spectral reflectance expression. With only four of the 28 comparisons with R values below 0.34 (albeit still above levels of the 999 R′ permutation values), the results inform us that the differences in the spectral reflectance expression between 24 of the 28 species comparisons allows successful species differentiation.

3.2.3. Wavelength Analysis to Find the Best Discriminating Wavelengths (SIMPER)

SIMPER analysis showed clear and distinct wavebands that are consistently contributing the most to dissimilarity between groups and species. For group differentiation, wavelengths of 535–570 nm dominate discrimination with additional narrow bands: 575–585 nm, 630–640 nm and 665–675 nm (Figure 6a). There are further wavebands that have contributed less, yet are still distinct and could be useful for enhancing the practical application of discriminating wavebands. Species-level SIMPER analysis revealed a single dominant waveband that consistently contributed the most across all 28 species comparisons; 570–590 nm with further areas of discrimination from 490–530 nm, and in the higher end of the spectrum from 610–620 nm, and 660–680 nm (Figure 6b). The wavelengths from 475–490 nm exhibit an area of particularly poor species discriminatory capability.

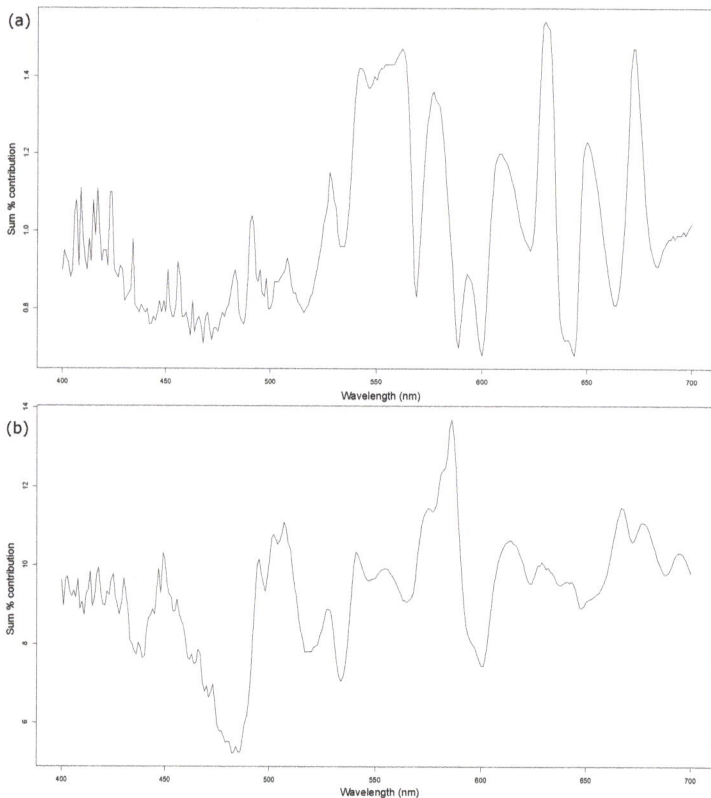

Figure 6. 1-way SIMPER analysis sum percentage contribution to significant dissimilarities made by wavelengths: (**a**) Between groups. (**b**) Between species.

4. Discussion

Coastal operators fight an ongoing battle with both vegetation and animal marine ingress entering water intakes. Most marine ingress occurrences around the UK arise from non-sessile macroalgae [60] and jellyfish [61] but can on occasions be caused by small shoals of fish [62]. Operators of desalination plants and nuclear power stations rely heavily on their water intakes to remain operational. If the water supply is interrupted, the plants must shut down. This results in the disruption of fresh water supply, or in the case of nuclear power stations, electricity export to the national grid. This is an issue that not only affects coastal operators, but the general public as well. Nuclear operators in particular require 6–8 h warning prior to marine ingress events occurring (EDF, personal comment). This warning helps to reduce significant losses in power generation and prevent more permanent damage. The work we report here focuses on the detection of non-sessile, low-density macroalgae that are found at various depths in the water column. It is this type of SAV that has caused continuing issues for UK nuclear power stations, with similar challenges being experienced in other countries as well. The overall aim of our work is to develop a regional-scale early-warning system for coastal operators to reduce their disruption and costs. An important factor in the development of such a system is to understand the reflectance signatures of the problematic macroalgae species. We aim to achieve this by identifying wavebands that will enable species to be distinguished from each other.

Investigating the spectral reflectance signature of each species is a critical step in deciding which wavelengths to incorporate within a remote sensing sensor for the identification of MABs [63]. The remote sensing of vegetation is usually highly dependent on the detection of reflected electromagnetic radiation [64]. Currently, it is only LiDAR and magnetometer sensors that do not adopt this approach [65]. Chlorophyllic vegetation, including seaweeds [66], have a characteristic signature which can be used to identify and discriminate it from other species [32,64,67]. Remote sensing of marine vegetation is restricted to wavelengths that can penetrate surface water, and secondly, to those that can be reflected to the sensor. The bands of the visible wavelength spectrum that can penetrate water effectively coincide with the areas of the spectrum that are most used by photosynthetic pigments found within chloroplasts [32]. Wavelengths above 700 nm begin to have reduced water penetration capability, only being able to pass through the upper layers of the water column [41,42,68]. Other factors such as organic and inorganic matter in the water column, phytoplankton presence, and surface water spectral scattering can cause further detection difficulties. For these reasons, it is likely that the remote sensing of MABs will require a high-resolution remote sensing system such as the centimeter spatial resolution imaging systems used by [69,70].

Throughout our data collection phase, we wanted to ensure that our collected spectra were as accurate and representative of the true profile as possible. Rather than taking measurements from a distance like [32,41], we wanted to ensure that we established a data set of robust lab-based reference measurements [71]. We also took all our readings in contact with the ASD measurement wand, which uses a controlled light source. By recording our spectra in this manner, we avoided the large sources of noise that the work by [72–74] had to overcome. Examples of such sources of noise are temporal variation, small instantaneous field of view imaging, undetected clouds, and poor atmospheric conditions. We also doubled the highest spectral averaging recommendation by [39] to produce naturally smooth spectral profiles (Figure 4) for all our readings. Unlike other examples of remote sensing that require smoothing filters [75–77] our data lack the sources of major noise that would normally demand the mandatory use of smoothing filters [73]. By taking an average of 379 readings per species, we were able to notably increase the power in our data set. This process also simultaneously accounted for vegetation variability within species. Figure 4b shows high-frequency noise around 400–450 nm; however, this is distinctly different to the major noise as previously mentioned and does not coincide with any SIMPER derived wavebands.

The pairwise heat map (Figure 5) was a valuable tool in facilitating a visual assessment of which parts of the spectrum could be used for species x species discrimination. Previous macroalgae research has generally been conducted using raw spectral readings [78], and not the 1st derivative, as has been used here. Our univariate investigation (Figure 5) found significance levels to be far higher than expected *a priori*. As found by [79], it is likely that the use of 1st derivative data is the source of this spectrum-wide increase in significance. The reasoning being due to its enhanced ability to highlight the signal [79]. Even for intra-group species comparisons, significance values, across the spectrum, were higher than expected. This again is likely to be due to the use of 1st derivative spectral data. The primary aim of the heat map was to visually identify bands of significance where the greatest number of comparisons could be differentiated. A total of 15 narrow (<10 nm) significant wavebands were identified across all 28 comparisons that could be used to differentiate between species, with the wavelengths from 550 to 750 nm being highly significant for 75% of all comparisons. The waveband of

570–590 nm (Figure 5) contains highly significant comparisons across all 28 comparisons, and precisely coincides with the optimal discriminatory waveband identified through the species-level SIMPER analysis (Figure 6b). This highly significant waveband is certainly due to pigmentation reflectance at these wavelengths [80] and can also be seen in Figure 4b.

The wavelengths between 550–750 nm show the most significance (Figure 5) across the majority of the 28 comparisons. This broad waveband has great potential for species discrimination. It exhibits high significance for most comparisons (23 out of a possible 28). Of the five comparisons that do not show high significance, four are intra-group comparisons with the other being *F. serratus* x *A. esculenta*. It would be reasonable to expect intra-group comparisons to have reduced significance, with respect to inter-group comparisons, as a result of morphological similarities. This would suggest that a sensor, tuned to detect the visible spectrum, would be most suitable to detect most comparisons. For the other 23 comparisons, there is minor non-significance within some of the ANOVA results (spanning 1–2 nm) but is unlikely to affect the practical use of this waveband. The source of this non-significance is uncertain, but is potentially due to minor fouling of the samples due to epiphytes and detritus [32,81]. Through the investigation and analysis of spectral reflectance signals of vegetation, we can identify which bands to target if the development of a remote sensing approach is to be successful [82]. By identifying the spectral reflectance characteristics of a photosynthetic species, it is even possible to detect individual conditions such as disease [83]. The most effective way of discriminating between species would be to differentiate species at a targeted pairwise level. However, this would not be practical when applied in a real-world remote sensing application. The species present at a particular location would not necessarily be known and therefore pairwise targeting would require extensive *a priori* species validation from the ground.

During investigation of the differences between the three sampled groups of kelp, fucoid, and other, the non-metric multidimensional scaling (Figure 3b) successfully demonstrated clear and distinct spectral differences with overlapping occurring for only a small proportion of the total readings. The species-level analysis (Figure 3a) allowed greater insight into the variability within the sampled groups. *U. lactuca* was the most spectrally distinct species analyzed being the sole member of the group other. There was significant spectral overlapping between the three fucoid species, and the four kelp species, albeit to a lesser extent. The data indicate that there are distinct spectral differences between the groups, but not between all species within a group. The visual similarity of the three fucoid species was noted prior to investigation and therefore the strong similarity in spectral reflectance expression, as shown in Figure 3a, is not surprising. The most distinct kelp clusterings of both *A. esculenta* and *L.* sp. stipe are to also be expected. *A. esculenta* being the only kelp species not belonging to the genus *Laminaria*, and *L.* sp. stipe being the only non-photosynthetic kelp species. The spread and spectral overlapping of both *L. saccharina* and *L.* sp. could be a result of there being some *L. saccharina* present in the *L.* sp. samples. Due to the geographical location of the sampling locations, it is probable that the *L.* sp. samples primarily consisted of two species—*L. hyperborea* and *L. digitata* with the addition of other species such as *L. saccharina*.

The reason Euclidean distances were used to produce the resemblance matrix was due to the presence of negative data values as a result of analyzing the 1st derivative spectra. This meant that it was not possible to use more commonly used dissimilarity scores, such as the Bray-Curtis statistic. However even after taking this constraint into account, it is unlikely that other similarity scoring methods would have yielded a different result. This is because of the superior power within the data set, compared to other well-known work [32], as a result of the large number of readings taken (Table 1). The results in Figure 3a suggest that species could be grouped with respect to their spectral similarity as follows: *U. lactuca*; *L.* sp. stipe; *L.* sp., *A. esculenta* & *L. saccharina*; and finally *F. serratus*, *F. spiralis* & *F. vesiculosus*.

ANOSIM analysis provided a more formal approach to the investigation of both inter-group, and intra-group, spectral reflectance. It must be noted that *p*-values for ANOSIM analyses are highly correlated with test power due to variation in sample sizes. Our focus was therefore on the stated

R values, which are an absolute measure of differences in spectral reflectance, with consideration of the *p*-value coming second. No pairwise comparison adjustment was used to maintain statistical transparency and to not provide a misrepresentation of certainty. The group ANOSIM analysis concluded that fucoid x other was the most spectrally distinct comparison (R = 0.712), with the kelp x other comparison being the most similar (R = 0.455) which is consistent with the findings of the nMDS (Figure 3b). This result suggests that despite the groups fucoid and other sharing similar ecological habitats—which could suggest the use of comparable photosynthetic pigments—this similarity has little impact on their overall spectral reflectance and distinctiveness.

There is an important requirement to monitor the extent and frequency of MABs to reduce their impact. A functioning remote sensing system could help to predict their arrival, thereby helping to protect high-value assets such as power stations in coastal locations. Due to the huge potential damage that can be caused by non-sessile low-density MABs [1,3,4], a way of predicting their movement in the form of an early-warning detection system would aid efforts to reduce their damaging effects [7]. The ability of a remote sensing system to distinguish between different species would be highly valuable. Different species can have various adverse impacts on coastal operators, with smaller species blocking water intakes while larger kelp species leading to impact and mechanical damages. The species-level ANOSIM analysis demonstrated that both *L.* sp. stipe and *U. lactuca* were the most spectrally distinct species and supported the visual perception. It is likely that the spectral dissimilarity exhibited by *L.* sp. stipe is due to it being the only species that lacks photosynthetic pigmentation. In contrast, the spectral dissimilarity shown by *U. lactuca* is likely due to it being the only representative of the group, other. We can conclusively differentiate 15 of all 28 potential comparisons while also being able to detect strong spectral differentiation for nine further comparisons even though there may be some minor similarities within specific comparisons. When taking into account that nine of the 28 comparisons are intra-group pairings, the outputs from the overall spectral analysis provide a firm foundation for developing a remote sensing capability for macroalgae in the marine environment. The ability to distinguish between the groups and to a species level for most comparisons is particularly useful.

It was possible to distinguish individual species between each of the three groups, but not necessarily within each group. The fucoids had the most similar spectral signatures yet *F. vesiculosus* was the only species, including the other kelp species, that shared detectable spectral similarity with *L.* sp. stipe. The vegetative structure of the stipes is significantly different from that of the photosynthetic kelp species.

There are generally high levels of broad-band reflectance of terrestrial vegetation in the near-infrared spectrum. This is predominantly due to internal leaf scattering offset by low levels of reflectance over 1300 nm due to strong wavelength absorption by water [20,32,33,41,64]. These characteristics are also present in most common British seaweeds. Much like a typical terrestrial plant, photosynthetic seaweeds have low reflectance within the visual spectrum due to chlorophyll absorption. This absorption occurs within the thylakoid sacs of the chloroplast [64]. However, this does not mean that the visual wavelength spectrum cannot be used to identify vegetation to species level. In fact, through the analysis of reflectance characteristics via wavelengths used for photosynthesis, it is well known that vegetation species can be successfully discriminated [84–86]. This is particularly relevant for early-warning detection systems that inherently require maximum water penetration; it is these lower wavebands that have superior water penetration capability [41,42,87].

Having detected significant levels of spectral differentiation between most species comparisons, this work has identified wavelengths that can be used in the design of a remote sensing methodology for the early detection of macroalgae ingress near nuclear power stations. The SIMPER analysis was particularly useful by simultaneously calculating the most representative wavelengths to use to identify species and to discriminate them from other species. This refined statistical approach enabled a single dominant waveband to be highlighted for species-level spectral differentiation, 570–590 nm (Figure 6b (sum % contribution = 14%)). Based on 1st derivative data, which by nature highlights the characteristics of the spectra and not the raw amplitudes, we can be confident that this waveband would be highly successful for species discrimination between our eight sampled species, and is supported by the ANOSIM analyses. The group-level SIMPER output was not as conclusive as it was to species level, where there was a single dominant peak. This is not an unexpected finding. Being able to determine a single optimum discriminatory waveband that can differentiate multiple species, nested within groups, is a particularly arduous task. However, this is not to say the process was unsuccessful. With a broad discriminatory waveband of 535–570 nm (sum % contribution of 1.4%) with three further narrow bands of differentiation above sum contributions of 1.2%, it is highly probable that effective group discrimination is possible. The primary discrimination band for group differentiation covers wavelengths previously known for their detection capabilities via chlorophyll pigmentation [16,32]. This would suggest that wavelengths associated with photosynthetic compounds are acceptable for group discrimination tasks. However, when requiring a more detailed species-level discrimination, wavelengths that are not associated with photosynthesis are more appropriate. We found that there was only enough variation to discriminate between the eight sampled species away from the chlorophyll associated wavebands, yet was still within the visible light spectrum.

For the provision of 6–8 h of warning prior to marine ingress events, we aim to focus on sensor types that can be fitted to UAV-based imaging systems. A regional-scale early-warning system using UAVs can provide solutions to the temporal, and atmospheric, challenges that satellite systems currently face. There are significant advances being made in UAV mountable sensor types [88,89] and are part of a rapidly advancing field of research [69,90,91]. Different sensor types can be tuned to specific parts of the spectrum using filters [92]; this is particularly common with charge-coupled devices (CCD) and CMOS sensors. These sensors are known for their sensitivity to the 400–1000 nm spectral range [92]. However, there can be sensor specific variations to the exact spectral range. Other examples of UAV mountable sensors include hyperspectral, thermal and LiDAR sensors with the latter two types showing great promise but are in the early stages of deployment on UAV platforms [91,93]. Hyperspectral sensors have been successfully used within agricultural surveying and have demonstrated the ability to collect high quality data [89]. However, this ability to collect high quality data has also become a challenge for their application to UAVs due to the resultant ortho-rectification errors [94]. Current airborne hyperspectral imaging also faces limitations from factors such as non-linear weather dynamics, irregular light intensity [94] as well as the weight of a survey grade sensor [88]. We do, however, agree with the findings of [88] that this is an extremely fast moving field and that there is great promise for drone-based hyperspectral imaging in the near future.

When applied to the practical discrimination of species, various imaging sensor techniques can be combined to improve overall image quality, but that does not necessarily result in improved species discrimination [95]. Our findings suggest that improved species discrimination can be more easily provided with a more selective waveband choice. With our identification of the 570–590 nm waveband for species discrimination, we recommend that a CCD-based sensor would be the most appropriate taking into account current limitations of other drone-scale sensors. CCDs are particularly sensitive to visible spectrum light, are lightweight, and easily mountable onto UAVs. The high-resolution capabilities of sensors fitted on UAVs [88,96], flexibility of sensor mounting options and their rapid deployment make them a prime candidate for the remote sensing of MABs with respect to coastal nuclear power stations as part of an early-warning detection system.

5. Conclusions

After sampling a total of eight macroalgae species, the use of 1st derivative spectral data was highly successful in identifying significant differences between both macroalgae groups, as well as species. In our univariate analysis, we identified that wavelengths of 570–590 nm had strong significance between all 28 comparisons. No broad wavebands (>30 nm) could differentiate all 28 comparisons. However, 15 narrow bands were identified that had high significance across all pairwise comparisons during the 1-way ANOVA pairwise analysis. Even though not belonging to the same genus, we found that *A. esculenta* and *L.* sp. had near identical spectral reflectance signatures.

During our multivariate analysis, we were able to successfully identify spectral differences between the three macroalgae groups, as well as for 100% of inter-group species comparisons. This contributed towards a species-level discriminatory success rate of 85% for all possible ANOSIM pairwise comparisons. We were not, however, able to differentiate between the three fucoid species.

Group differentiation was found to be associated with chlorophyll pigmentation (535–570 nm) while the more demanding task of species differentiation was accomplished with a waveband (570–590 nm) away from wavelengths strongly associated with chlorophyll. During our SIMPER analyses, this single dominant waveband (570–590 nm) was identified as a consistent contributor to the differentiation of all eight species. This is consistent with the key output of our univariate analysis. The use of this waveband is recommended for further investigation and the practical testing of it for real-world species discrimination. We will now investigate the use of a UAV mounted CCD-based sensor focused on the 570–590 nm waveband that was identified, as the next phase in the development of a regional-scale early-warning detection system for potentially disruptive MABs.

Author Contributions: B.M. Wrote the paper, conceived and designed the investigation and methodology, curated and analyzed the data, and completed the visualization. B.M. & M.R.C. completed the validation. M.R.C. provided the resources and was the supervisor for the project. M.R.C. & P.L. acquired the project funding, provided project administration, and reviewed and edited the final manuscript.

Funding: This research was funded by the Engineering and Physical Sciences Research Council (EPSRC), The Smith Institute and EDF Energy under EPSRC Industrial Case Studentship Voucher 586 Number 16000001.

Acknowledgments: We would like to thank the Engineering and Physical Sciences Research Council (EPSRC), The Smith Institute and EDF Energy for funding this project. We would also like to thank the reviewers for their helpful comments and constructive criticism. The manuscript became stronger thanks to their detailed contribution. The underlying data are confidential and cannot be shared.

Appendix A

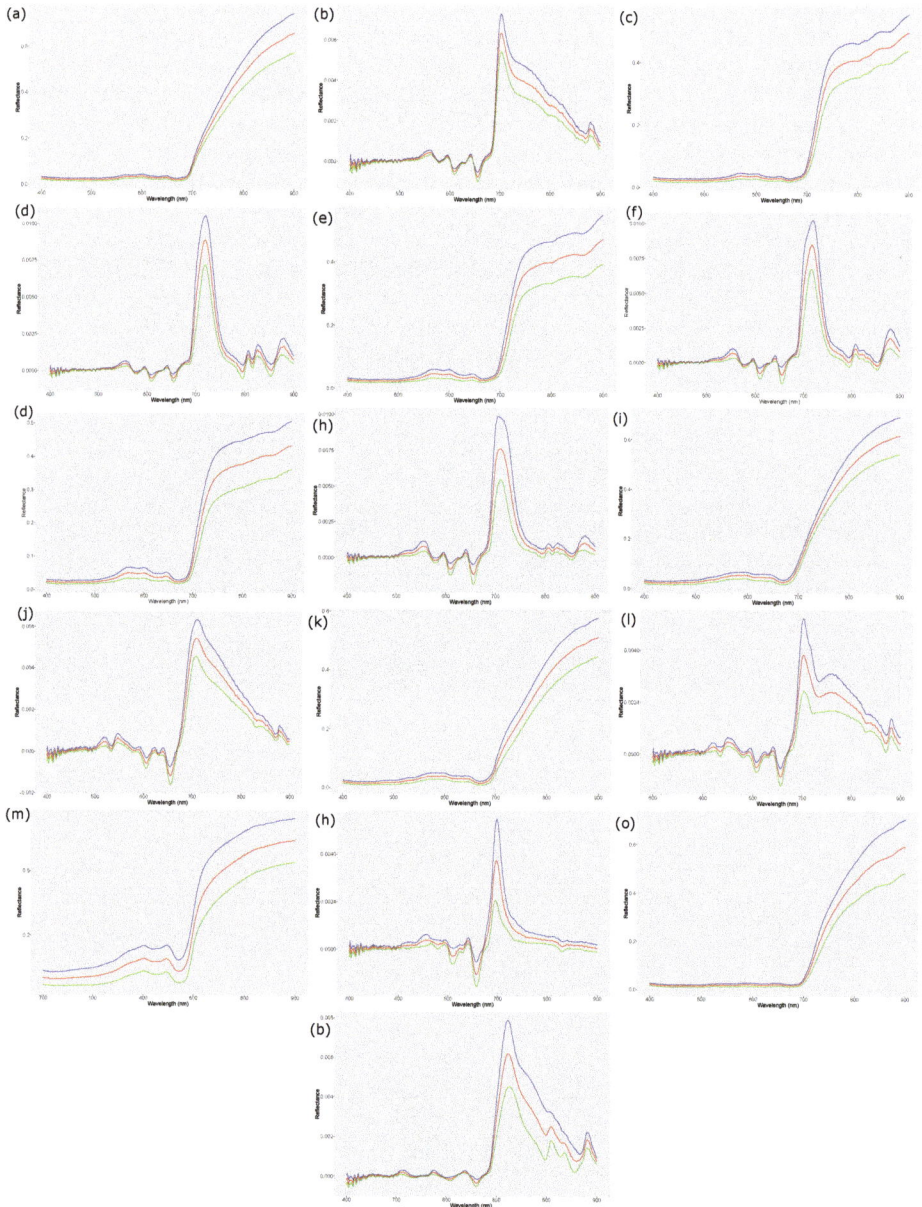

Figure A1. Spectral outputs for each species, red = mean, blue = mean + 1SD, green = mean −
1SD: (**a**) *Alaria esculenta* raw. (**b**) *Alaria esculenta* 1st derivative. (**c**) *Fucus serratus* raw. (**d**) *Fucus
serratus* 1st derivative. (**e**) *Fucus spiralis* raw. (**f**) *Fucus spiralis* 1st derivative. (**g**) *Fucus vesiculosus* raw.
(**h**) *Fucus vesiculosus* 1st derivative. (**i**) *Laminaria saccharina* raw. (**j**) *Laminaria saccharina* 1st derivative.
(**k**) *Laminaria sp* raw. (**l**) *Laminaria sp* 1st derivative. (**m**) *Laminaria sp stipe* raw. (**n**) *Laminaria sp stipe* 1st
derivative. (**o**) *Ulva lactuca* raw. (**p**) *Ulva lactuca* 1st derivative.

Remote Sens. **2019**, *11*, 704

References

1. Lapointe, B.E.; Bedford, B.J. Drift rhodophyte blooms emerge in Lee County, Florida, USA: Evidence of escalating coastal eutrophication. *Harmful Algae* **2007**, *6*, 421–437. [CrossRef]
2. Brand, L.E.; Compton, A. Long-term increase in Karenia brevis abundance along the Southwest Florida Coast. *Harmful Algae* **2007**, *6*, 232–252. [CrossRef] [PubMed]
3. Fletcher, R.L. *The Occurrence of "Green Tides"—A Review*; Springer: Berlin, Germany, 1996; pp. 7–43. [CrossRef]
4. Ye, N.H.; Zhang, X.W.; Mao, Y.Z.; Liang, C.W.; Xu, D.; Zou, J.; Zhuang, Z.M.; Wang, Q.Y. 'Green tides' are overwhelming the coastline of our blue planet: Taking the world's largest example. *Ecol. Res.* **2011**, *26*, 477–485. [CrossRef]
5. Glibert, P.; Seitzinger, S.; Heil, C.; Burkholder, J.; Parrow, M.; Codispoti, L.; Kelly, V. The Role of Eutrophication in the Global Proliferation of Harmful Algal Blooms. *Oceanography* **2005**, *18*, 198–209. [CrossRef]
6. Howarth, R.W.; Sharpley, A.; Walker, D. Sources of nutrient pollution to coastal waters in the United States: Implications for achieving coastal water quality goals. *Estuaries* **2002**, *25*, 656–676. [CrossRef]
7. Pang, S.J.; Liu, F.; Shan, T.F.; Xu, N.; Zhang, Z.H.; Gao, S.Q.; Chopin, T.; Sun, S. Tracking the algal origin of the Ulva bloom in the Yellow Sea by a combination of molecular, morphological and physiological analyses. *Mar. Environ. Res.* **2010**, *69*, 207–215. [CrossRef] [PubMed]
8. Cox, P.A.; Banack, S.A.; Murch, S.J. Biomagnification of cyanobacterial neurotoxins and neurodegenerative disease among the Chamorro people of Guam. *Proc. Natl. Acad. Sci. USA* **2003**, *100*, 13380–13383. [CrossRef]
9. De Vries, I.; Philippart, C.J.M.; DeGroodt, E.G.; van der Tol, M.W.M. *Coastal Eutrophication and Marine Benthic Vegetation: A Model Analysis*; Springer: Berlin/Heidelberg, Germany, 1996; pp. 79–113. [CrossRef]
10. Shen, H.; Perrie, W.; Liu, Q.; He, Y. Detection of macroalgae blooms by complex SAR imagery. *Mar. Pollut. Bull.* **2014**, *78*, 190–195. [CrossRef] [PubMed]
11. Villacorte, L.; Tabatabai, S.; Dhakal, N.; Amy, G.; Schippers, J.; Kennedy, M. Algal blooms: An emerging threat to seawater reverse osmosis desalination. *Desalin. Water Treat.* **2015**, *55*, 2601–2611. [CrossRef]
12. Hallegraeff, G.M. A review of harmful algal blooms and their apparent global increase. *Phycologia* **1993**, *32*, 79–99. [CrossRef]
13. Zingone, A.; Oksfeldt Enevoldsen, H. The diversity of harmful algal blooms: A challenge for science and management. *Ocean Coast. Manag.* **2000**, *43*, 725–748. [CrossRef]
14. Cai, T.; Park, S.Y.; Li, Y. Nutrient recovery from wastewater streams by microalgae: Status and prospects. *Renew. Sustain. Energy Rev.* **2013**, *19*, 360–369. [CrossRef]
15. Bonanno, G.; Orlando-Bonaca, M. Trace elements in Mediterranean seagrasses and macroalgae. A review. *Sci. Total Environ.* **2018**, *618*, 1152–1159. [CrossRef] [PubMed]
16. Gitelson, A.A.; Merzlyak, M.N. Signature Analysis of Leaf Reflectance Spectra: Algorithm Development for Remote Sensing of Chlorophyll. *J. Plant Physiol.* **1996**, *148*, 494–500. [CrossRef]
17. Dekker, A.G.; Phinn, S.R.; Anstee, J.; Bissett, P.; Brando, V.E.; Casey, B.; Fearns, P.; Hedley, J.; Klonowski, W.; Lee, Z.P.; et al. Intercomparison of Shallow Water Bathymetry, Hydro-Optics, and Benthos Mapping Techniques in Australian and Caribbean Coastal Environments. *Limnol. Oceanogr.* **2011**, *9*, 396–425. [CrossRef]
18. Hedley, J.; Enríquez, S. Optical properties of canopies of the tropical seagrass Thalassia testudinum estimated by a three-dimensional radiative transfer model. *Limnol. Oceanogr.* **2010**. [CrossRef]
19. Hedley, J.; Russell, B.; Randolph, K.; Dierssen, H. A physics-based method for the remote sensing of seagrasses. *Remote Sens. Environ.* **2016**, *174*, 134–147. [CrossRef]
20. Dierssen, H.M.; Zimmerman, R.C.; Leathers, R.A.; Downes, T.V.; Davis, C.O. Ocean color remote sensing of seagrass and bathymetry in the Bahamas Banks by high-resolution airborne imagery. *Limnol. Oceanogr.* **2003**, *48*, 444–455. [CrossRef]
21. Hu, C.; Feng, L.; Hardy, R.F. Spectral and spatial requirements of remote measurements of pelagic Sargassum macroalgae. *Remote Sens. Environ.* **2015**, *167*, 229–246. [CrossRef]
22. Parsiani, H.; Torres, M.; Rodriguez, P.A. High-resolution vegetation index as measured by radar and its validation with spectrometer. In Proceedings of the Image and Signal Processing for Remote Sensing X, Canary Islands, Spain, 13–16 September 2004; p. 284. [CrossRef]

23. Zhang, X.; Kondragunta, S.; Ram, J.; Schmidt, C.; Huang, H.C.; Zhang, X. Near-real-time global biomass burning emissions product from geostationary satellite constellation. *J. Geophys. Res.* **2012**, *117*, D14. [CrossRef]

24. Saunders, C.; Bird, R.; Da Silva, A.; Sweeting, M.; Gomes, L. Design considerations in rapid-revisit small satellite constellations. In Proceedings of the 68th International Astronautical Congress: Unlocking Imagination, Fostering Innovation and Strengthening Security, Adelaide, Australia, 25–29 September 2017; pp. 5961–5984.

25. Lubac, B.; Loisel, H. Variability and classification of remote sensing reflectance spectra in the eastern English Channel and southern North Sea. *Remote Sens. Environ.* **2007**, *110*, 45–58. [CrossRef]

26. Zou, W.; Yuan, L.; Zhang, L. Analyzing the spectral response of submerged aquatic vegetation in a eutrophic lake, Shanghai, China. *Ecol. Eng.* **2013**, *57*, 65–71. [CrossRef]

27. Huete, A.; Didan, K.; Miura, T.; Rodriguez, E.; Gao, X.; Ferreira, L. Overview of the radiometric and biophysical performance of the MODIS vegetation indices. *Remote Sens. Environ.* **2002**, *83*, 195–213. [CrossRef]

28. Fassnacht, F.E.; Latifi, H.; Stereńczak, K.; Modzelewska, A.; Lefsky, M.; Waser, L.T.; Straub, C.; Ghosh, A. Review of studies on tree species classification from remotely sensed data. *Remote Sens. Environ.* **2016**, *186*, 64–87. [CrossRef]

29. Längkvist, M.; Kiselev, A.; Alirezaie, M.; Loutfi, A. Classification and Segmentation of Satellite Orthoimagery Using Convolutional Neural Networks. *Remote Sens.* **2016**, *8*, 329. [CrossRef]

30. Pérez-Ortiz, M.; Peña, J.M.; Gutiérrez, P.A.; Torres-Sánchez, J.; Hervás-Martínez, C.; López-Granados, F. Selecting patterns and features for between- and within- crop-row weed mapping using UAV-imagery. *Expert Syst. Appl.* **2016**, *47*, 85–94. [CrossRef]

31. Tin, H.C.; O'Leary, M.; Fotedar, R.; Garcia, R. Spectral response of marine submerged aquatic vegetation: A case study in Western Australia coast. In Proceedings of the OCEANS 2015—MTS/IEEE Washington, Washington, DC, USA, 19–22 October 2015; pp. 1–5. [CrossRef]

32. Fyfe, S.K. Spatial and temporal variation in spectral reflectance: Are seagrass species spectrally distinct? *Limnol. Oceanogr.* **2003**, *48*, 464–479. [CrossRef]

33. Hu, L.; Hu, C.; Ming-Xia, H. Remote estimation of biomass of Ulva prolifera macroalgae in the Yellow Sea. *Remote Sens. Environ.* **2017**, *192*, 217–227. [CrossRef]

34. Xing, Q.; Hu, C. Mapping macroalgal blooms in the Yellow Sea and East China Sea using HJ-1 and Landsat data: Application of a virtual baseline reflectance height technique. *Remote Sens. Environ.* **2016**, *178*, 113–126. [CrossRef]

35. Gao, Y.; Fang, J.; Zhang, J.; Ren, L.; Mao, Y.; Li, B.; Zhang, M.; Liu, D.; Du, M. The impact of the herbicide atrazine on growth and photosynthesis of seagrass, *Zostera marina* (L.), seedlings. *Mar. Pollut. Bull.* **2011**, *62*, 1628–1631. [CrossRef]

36. Simic, A.; Chen, J.M. Refining a hyperspectral and multiangle measurement concept for vegetation structure assessment. *Can. J. Remote Sens.* **2008**, *34*, 174–191. [CrossRef]

37. Nuclear Energy Institute. *Economic Impacts of The R.E. Ginna Nuclear Power Plant an Analysis*; Technical Report; Nuclear Energy Institute: Washington, DC, USA, 2015.

38. Labsphere. *Spectralon® Diffuse Reflectance Standards*; Labsphere | Internationally Recognized Photonics Company: North Sutton, NH, USA, 2018.

39. Danner, M.; Locherer, M.; Hank, T.; Richter, K. *Spectral Sampling with the ASD FieldSpec 4*; GFZ Data Services: Potsdam, Germany, 2015. [CrossRef]

40. Fabre, S.; Lesaignoux, A.; Olioso, A.; Briottet, X. Influence of Water Content on Spectral Reflectance of Leaves in the 3–15 μm Domain. *IEEE Geosci. Remote Sens. Lett.* **2011**, *8*, 143–147. [CrossRef]

41. Garcia, R.; Hedley, J.; Tin, H.; Fearns, P.; Garcia, R.A.; Hedley, J.D.; Tin, H.C.; Fearns, P.R.C.S. A Method to Analyze the Potential of Optical Remote Sensing for Benthic Habitat Mapping. *Remote Sens.* **2015**, *7*, 13157–13189. [CrossRef]

42. Kirk, J.T.O. *Light and Photosynthesis in Aquatic Ecosystems*; Cambridge University Press: Cambridge, UK, 1994; p. 509.

43. ASD. *Indico Pro*; Malvern Panalytical: Malvern, UK, 2017.

44. Hong, Y.; Liu, Y.; Chen, Y.; Liu, Y.; Yu, L.; Liu, Y.; Cheng, H. Application of fractional-order derivative in the quantitative estimation of soil organic matter content through visible and near-infrared spectroscopy. *Geoderma* **2019**, *337*, 758–769. [CrossRef]

45. Fligner, M.A.; Killeen, T.J. Distribution-Free Two-Sample Tests for Scale. *J. Am. Stat. Assoc.* **1976**, *71*, 210–213. [CrossRef]

46. Kruskal, W.H.; Wallis, W.A. Use of Ranks in One-Criterion Variance Analysis. *J. Am. Stat. Assoc.* **1952**, *47*, 583–621. [CrossRef]

47. Bartlett, M.S. A Note on the Multiplying Factors for Various χ^2 Approximations. *J. R. Stat. Soc.* **1954**. [CrossRef]

48. Delacre, M.; Lakens, D.; Leys, C. Why Psychologists Should by Default Use Welch's *t*-test Instead of Student's *t*-test. *Int. Rev. Soc. Psychol.* **2017**, *30*, 92. [CrossRef]

49. Jaccard, J.; Becker, M.A.; Wood, G. Pairwise multiple comparison procedures: A review. *Psychol. Bull.* **1984**, *96*, 589–596. [CrossRef]

50. Kolde, R. Pheatmap: Pretty Heatmaps. 2018. Available online: https://cran.mtu.edu/web/packages/pheatmap/index.html (accessed on 28 January 2019).

51. R Core Team. R: A Language and Environment for Statistical Computing. 2017. Available online: https://www.r-project.org/ (accessed on 28 January 2019).

52. Oksanen, J.; Guillaume Blanchet, F.; Friendly, M.; Kindt, R.; Legendre, P.; McGlinn, D.; Michin, P.; O'Hara, R.; Simpson, G.; Solymos, P.; et al. Vegan: Community Ecology Package. 2018. Available online: https://CRAN.R-project.org/package=vegan (accessed on 28 January 2019).

53. Wickham, H. *ggplot2: Elegant Graphics for Data Analysis*; Springer: New York, NY, USA, 2009.

54. Wickham, H. Reshaping Data with the reshape Package. *J. Stat. Softw.* **2007**, *21*, 1–20. [CrossRef]

55. Clarke, R.; Gorley, R. PRIMER-E. 2015. Available online: http://www.primer-e.com/ (accessed on 28 January 2019).

56. Clarke, K.R. Non-parametric multivariate analyses of changes in community structure. *Aust. J. Ecol.* **1993**, *18*, 117–143. [CrossRef]

57. Buttigieg, P.L.; Ramette, A. A guide to statistical analysis in microbial ecology: A community-focused, living review of multivariate data analyses. *FEMS Microbiol. Ecol.* **2014**, *90*, 543–550. [CrossRef] [PubMed]

58. Baeck, P.; Blommaert, J.; Delalieux, S.; Delauré, B.; Livens, S.; Nuyts, D.; Sima, A.; Jacquemin, G.; Goffart, J.P.; Nv, V.; et al. High resolution vegetation mapping with a novel compact hyperspectral camera system. In Proceedings of the 13th International Conference on Precision Agriculture, St. Louis, MO, USA, 31 July–4 August 2016.

59. Wells, E. *A Field Guide to the British Seaweeds as Required for Assistance in the Classification of Water Bodies under the Water Framework Directive*; Environment Agency: Bristol, UK, 1997.

60. Vaughan, A. In a Laver: Seaweed Shuts Nuclear Reactor Again in Bad Weather. *The Guardian*, 5 March 2018.

61. Matsumura, K.; Kamiya, K.; Yamashita, K.; Hayashi, F.; Watanabe, I.; Murao, Y.; Miyasaka, H.; Kamimura, N.; Nogami, M. Genetic polymorphism of the adult medusae invading an electric power station and wild polyps of Aurelia aurita in Wakasa Bay, Japan. *J. Mar. Biol. Assoc. UK* **2005**, *85*, 563–568. [CrossRef]

62. Barath Kumar, S.; Mohanty, A.; Das, N.; Satpathy, K.; Sarkar, S. Impingement of marine organisms in a tropical atomic power plant cooling water system. *Mar. Pollut. Bull.* **2017**, *124*, 555–562. [CrossRef] [PubMed]

63. Dekker, A.G.; Malthus, T.J.; Wijnen, M.M.; Seyhan, E. Remote sensing as a tool for assessing water quality in Loosdrecht lakes. *Hydrobiologia* **1992**, *233*, 137–159. [CrossRef]

64. Knipling, E.B. Physical and physiological basis for the reflectance of visible and near-infrared radiation from vegetation. *Remote Sens. Environ.* **1970**, *1*, 155–159. [CrossRef]

65. Clothiaux, E.E.; Ackerman, T.P.; Mace, G.G.; Moran, K.P.; Marchand, R.T.; Miller, M.A.; Martner, B.E. Objective Determination of Cloud Heights and Radar Reflectivities Using a Combination of Active Remote Sensors at the ARM CART Sites. *J. Appl. Meteorol.* **2000**, *39*, 645–665. [CrossRef]

66. Suwandana, E.; Kawamura, K.; Sakuno, Y.; Evri, M.; Lesmana, A.H. Hyperspectral Reflectance Response of Seagrass (*Enhalus acoroides*) and Brown Algae (*Sargassum* sp.) to Nutrient Enrichment at Laboratory Scale. *J. Coast. Res.* **2012**, *283*, 956–963. [CrossRef]

67. He, K.S.; Rocchini, D.; Neteler, M.; Nagendra, H. Benefits of hyperspectral remote sensing for tracking plant invasions. *Divers. Distrib.* **2011**, *17*, 381–392. [CrossRef]

68. Jacques, S.L. Optical properties of biological tissues: A review. *Phys. Med. Biol.* **2013**, *58*, R37–R61. [CrossRef]

69. Carrasco-Escobar, G.; Manrique, E.; Ruiz-Cabrejos, J.; Saavedra, M.; Alava, F.; Bickersmith, S.; Prussing, C.; Vinetz, J.M.; Conn, J.E.; Moreno, M.; et al. High-accuracy detection of malaria vector larval habitats using drone-based multispectral imagery. *PLoS Negl. Trop. Dis.* **2019**, *13*, e0007105. [CrossRef]

70. Jay, S.; Baret, F.; Dutartre, D.; Malatesta, G.; Héno, S.; Comar, A.; Weiss, M.; Maupas, F. Exploiting the centimeter resolution of UAV multispectral imagery to improve remote-sensing estimates of canopy structure and biochemistry in sugar beet crops. *Remote Sens. Environ.* **2018**. [CrossRef]

71. Atzberger, C.; Eilers, P.H.C. Evaluating the effectiveness of smoothing algorithms in the absence of ground reference measurements. *Int. J. Remote Sens.* **2011**, *32*, 3689–3709. [CrossRef]

72. Atzberger, C.; Eilers, P.H. A time series for monitoring vegetation activity and phenology at 10-daily time steps covering large parts of South America. *Int. J. Digit. Earth* **2011**, *4*, 365–386. [CrossRef]

73. Atzberger, C.; Wess, M.; Doneus, M.; Verhoeven, G. ARCTIS—A MATLAB® Toolbox for Archaeological Imaging Spectroscopy. *Remote Sens.* **2014**, *6*, 8617–8638. [CrossRef]

74. Meroni, M.; Fasbender, D.; Rembold, F.; Atzberger, C.; Klisch, A. Near real-time vegetation anomaly detection with MODIS NDVI: Timeliness vs. accuracy and effect of anomaly computation options. *Remote Sens. Environ.* **2019**, *221*, 508–521. [CrossRef] [PubMed]

75. Atzberger, C.; Klisch, A.; Mattiuzzi, M.; Vuolo, F.; Atzberger, C.; Klisch, A.; Mattiuzzi, M.; Vuolo, F. Phenological Metrics Derived over the European Continent from NDVI3g Data and MODIS Time Series. *Remote Sens.* **2013**, *6*, 257–284. [CrossRef]

76. Doneus, M.; Verhoeven, G.; Atzberger, C.; Wess, M.; Ruš, M. New ways to extract archaeological information from hyperspectral pixels. *J. Archaeol. Sci.* **2014**, *52*, 84–96. [CrossRef]

77. Rembold, F.; Atzberger, C.; Savin, I.; Rojas, O.; Rembold, F.; Atzberger, C.; Savin, I.; Rojas, O. Using Low Resolution Satellite Imagery for Yield Prediction and Yield Anomaly Detection. *Remote Sens.* **2013**, *5*, 1704–1733. [CrossRef]

78. Zacharias, M.; Niemann, O.; Borstad, G. An Assessment and Classification of a Multispectral Bandset for the Remote Sensing of Intertidal Seaweeds. *Can. J. Remote Sens.* **1992**, *18*, 263–274. [CrossRef]

79. Gong, P.; Pu, R.; Yu, B. Conifer species recognition: An exploratory analysis of In Situ Hyperspectral data. *Remote Sens. Environ.* **1997**, *62*, 189–200. [CrossRef]

80. Carter, G.A.; Knapp, A.K. Leaf optical properties in higher plants: Linking spectral characteristics to stress and chlorophyll concentration; Leaf optical properties in higher plants: Linking spectral characteristics to stress and chlorophyll concentration. *Am. J. Bot.* **2001**. [CrossRef]

81. Wernberg, T.; Krumhansl, K.; Filbee-Dexter, K.; Pedersen, M.F. Status and Trends for the World's Kelp Forests. In *World Seas: An Environmental Evaluation*; Academic Press: Cambridge, MA, USA, 2019; pp. 57–78. [CrossRef]

82. Sridhar, B.B.M.; Han, F.X.; Diehl, S.V.; Monts, D.L.; Su, Y. Spectral reflectance and leaf internal structure changes of barley plants due to phytoextraction of zinc and cadmium. *Int. J. Remote Sens.* **2007**, *28*, 1041–1054, doi:10.1080/01431160500075832. [CrossRef]

83. Pacumbaba, R.; Beyl, C. Changes in hyperspectral reflectance signatures of lettuce leaves in response to macronutrient deficiencies. *Adv. Space Res.* **2011**, *48*, 32–42. [CrossRef]

84. Datt, B. Recognition of Eucalyptus forest species using hyperspectral reflectance data. In Proceedings of the IEEE 2000 International Geoscience and Remote Sensing Symposium. Taking the Pulse of the Planet: The·Role of Remote Sensing in Managing the Environment, Honolulu, HI, USA, 24–28 July 2000; Volume 4, pp. 1405–1407. [CrossRef]

85. Sims, D.A.; Gamon, J.A. Relationships between leaf pigment content and spectral reflectance across a wide range of species, leaf structures and developmental stages. *Remote Sens. Environ.* **2002**, *81*, 337–354. [CrossRef]

86. Schmidt, K.; Skidmore, A. Spectral discrimination of vegetation types in a coastal wetland. *Remote Sens. Environ.* **2003**, *85*, 92–108. [CrossRef]

87. Morris, W.D.; Witte, W.G.; Whitlock, C.H. *Turbid Water Measurements of Remote Sensing Penetration Depth at Visible and Near-Infrared Wavelength*; Technical Report; NASA—Langley Research Center: Hampton, VA, USA, 1980.

88. Aasen, H.; Honkavaara, E.; Lucieer, A.; Zarco-Tejada, P.; Aasen, H.; Honkavaara, E.; Lucieer, A.; Zarco-Tejada, P.J. Quantitative Remote Sensing at Ultra-High Resolution with UAV Spectroscopy: A Review of Sensor Technology, Measurement Procedures, and Data Correction Workflows. *Remote Sens.* **2018**, *10*, 1091. [CrossRef]

89. Greenwood, W.W.; Lynch, J.P.; Zekkos, D. Applications of UAVs in Civil Infrastructure. *J. Infrastruct. Syst.* **2019**, *25*, 04019002. [CrossRef]

90. Gray, P.; Ridge, J.; Poulin, S.; Seymour, A.; Schwantes, A.; Swenson, J.; Johnston, D.; Gray, P.C.; Ridge, J.T.; Poulin, S.K.; et al. Integrating Drone Imagery into High Resolution Satellite Remote Sensing Assessments of Estuarine Environments. *Remote Sens.* **2018**, *10*, 1257. [CrossRef]

91. Kays, R.; Sheppard, J.; Mclean, K.; Welch, C.; Paunescu, C.; Wang, V.; Kravit, G.; Crofoot, M. Hot monkey, cold reality: Surveying rainforest canopy mammals using drone-mounted thermal infrared sensors. *Int. J. Remote Sens.* **2019**, *40*, 407–419. [CrossRef]

92. Yang, C. A high-resolution airborne four-camera imaging system for agricultural remote sensing. *Comput. Electron. Agric.* **2012**, *88*, 13–24. [CrossRef]

93. Tang, L.; Shao, G. Drone remote sensing for forestry research and practices. *J. For. Res.* **2015**, *26*, 791–797. [CrossRef]

94. Singh, K.D.; Nansen, C. Advanced calibration to improve robustness of drone-acquired hyperspectral remote sensing data. In Proceedings of the 2017 6th International Conference on Agro-Geoinformatics, Fairfax, VA, USA, 7–10 August 2017; pp. 1–6. [CrossRef]

95. Bostater, C.R.; Jones, J.; Frystacky, H.; Kovacs, M.; Jozsa, O. Integration, testing, and calibration of imaging systems for land and water remote sensing. In Proceedings of the Remote Sensing of the Ocean, Sea Ice, and Large Water Regions 2010, Toulouse, France, 20–23 September 2010; Volume 7825, p. 78250N. [CrossRef]

96. Larson, M.D.; Simic Milas, A.; Vincent, R.K.; Evans, J.E. Multi-depth suspended sediment estimation using high-resolution remote-sensing UAV in Maumee River, Ohio. *Int. J. Remote Sens.* **2018**, *39*, 5472–5489. [CrossRef]

remote sensing

MDPI

Article

Performance Evaluation of Newly Proposed Seaweed Enhancing Index (SEI)

Muhammad Danish Siddiqui [1], Arjumand Z. Zaidi [1,2,*] and Muhammad Abdullah [1]

[1]   Department of Remote Sensing and Geoinformation Science, Institute of Space Technology, Karachi 75270, Pakistan; mdanish.ncrg04@ist.edu.pk (M.D.S.); abdullah.ncrg04@ist.edu.pk (M.A.)
[2]   US Pakistan Centers for Advanced Studies in Water, Mehran University of Engineering and Technology, Jamshoro, Sindh 76062, Pakistan
*   Correspondence: arjumand.uspcasw@faculty.muet.edu.pk

Received: 17 May 2019; Accepted: 27 May 2019; Published: 17 June 2019

Abstract: Seaweed is a valuable coastal resource for its use in food, cosmetics, and other items. This study proposed new remote sensing based seaweed enhancing index (SEI) using spectral bands of near-infrared (NIR) and shortwave-infrared (SWIR) of Landsat 8 satellite data. Nine Landsat 8 satellite images of years 2014, 2016, and 2018 for the January, February, and March months were utilized to test the performance of SEI. The seaweed patches in the coastal waters of Karachi, Pakistan were mapped using the SEI, normalized difference vegetation index (NDVI), and floating algae index (FAI). Seaweed locations recorded during a field survey on February 26, 2014, were used to determine threshold values for all three indices. The accuracy of SEI was compared with NDVI while placing FAI as the reference index. The accuracy of NDVI and SEI were assessed by matching their spatial extent of seaweed cover with FAI enhanced seaweed area. SEI images of January 2016, February 2018, and March 2018 enhanced less than 50 percent of the corresponding FAI total seaweed areas. However, on these dates the NDVI performed very well, matching more than 95 percent of FAI seaweed coverage. Except for these three times, the performance of SEI in the remaining six images was either similar to NDVI or even better than NDVI. SEI enhanced 99 percent of FAI seaweed cover on January 2018 image. Overall, seaweed area not covered by FAI was greater in SEI than NDVI in almost all images, which needs to be further explored in future studies by collecting extensive field information to validate SEI mapped additional area beyond the extent of FAI seaweed cover. Based on these results, in the majority of the satellite temporal images selected for this study, the performance of the newly proposed index—SEI, was found either better than or similar to NDVI.

Keywords: floating algae index (FAI); normalized difference vegetation index (NDVI); remote sensing; seaweed enhancing index (SEI); seaweed

1. Introduction

Seaweed is the name given to the numerous marine plants and algae that animate in seas, oceans, rivers, lakes, and other water forms. Seaweeds can be of three types based on the pigments they contain [1]. Their light-absorbing pigments can be either red, green, or brown. Depending upon these pigments, seaweeds perform their process of photosynthesis. Seaweed stock is an important component of the coastal ecosystem that provides living space for mangroves and coral reefs and breeding grounds and food for several types of nearshore fish, shrimp, marine reptiles, shellfish, and mammals [2,3]. Seaweeds also purify water for fish aquaculture. In recent years, human activities have impacted seaweed biodiversity by destroying seaweed habitat mainly caused by coastal pollution [4]. The beneficial chemical properties and nutritional value of seaweed have made it a commercially important coastal product. Generally, it is consumed in many countries of the world as human food, livestock fodder, and agricultural fertilizer [5]. During the years 2000 to 2014, global seaweed production was doubled from 10.5 to 28.4 million tonnes. The world's seaweed production in 2012 was estimated at around US$6 billion, and 95 percent of this production was from Asian countries' aquaculture [6].

Seaweed resources are present along the Pakistan coast. Seventy different classes and twenty-seven different categories of seaweed are reported from the coastal areas of Pakistan, *Ulva fasiata*, *Chondria tennussima*, *Sargassum* spp, and *Valoniopsis pachynema* are the most richly found species of seaweeds in this region [7]. Despite having great economic potential, these natural resources are still unmapped and unexplored. The reason is mainly the lack of monitoring and conservation endeavors in the country. Another reason might be the general ignorance about its environmental importance and economic potential. At present, seaweed is not used at a large scale in Pakistan as a consumable item. To fully utilize seaweeds' economic potential, it is necessary to explore and map seaweed stock that is available in Pakistan. Mapping and monitoring of seaweed and other benthic feature are needed to protect these natural resources.

Benthic maps are significant for management, research, and planning of marine resources. Mapping seaweed resources covering larger spatial areas using conventional methods through field investigations are capital intensive and time-consuming. Remote sensing (RS) is a useful tool for observing benthic habitats such as benthic algae and coral-reef ecosystems. For thematic mapping, habitats are defined as spatially distinct areas where physical, chemical, and the biological characteristics are distinctively different from nearby regions [8]. Satellite remote sensing can provide timely and updated information for monitoring high spatial and temporal variations of coastal resources, including seaweed stocks [9]. Numerous researchers have tested airborne and spaceborne sensor systems for marine studies [10]. The present study was undertaken to map seaweed resources along the Karachi Coast using geospatial techniques.

2. Material and Methods

2.1. Study Area and Satellite Data

Seaweed resources in Pakistan are still unmapped. The study sites for the present work are located offshore the Hawks Bay beach along the Karachi Coast, Sindh. These sites were selected through preliminary boat survey, which was conducted during February 2014. GPS points were recorded on seaweed patches and overlaid on the satellite imagery of the same date (February 26, 2014) and location (Figure 1).

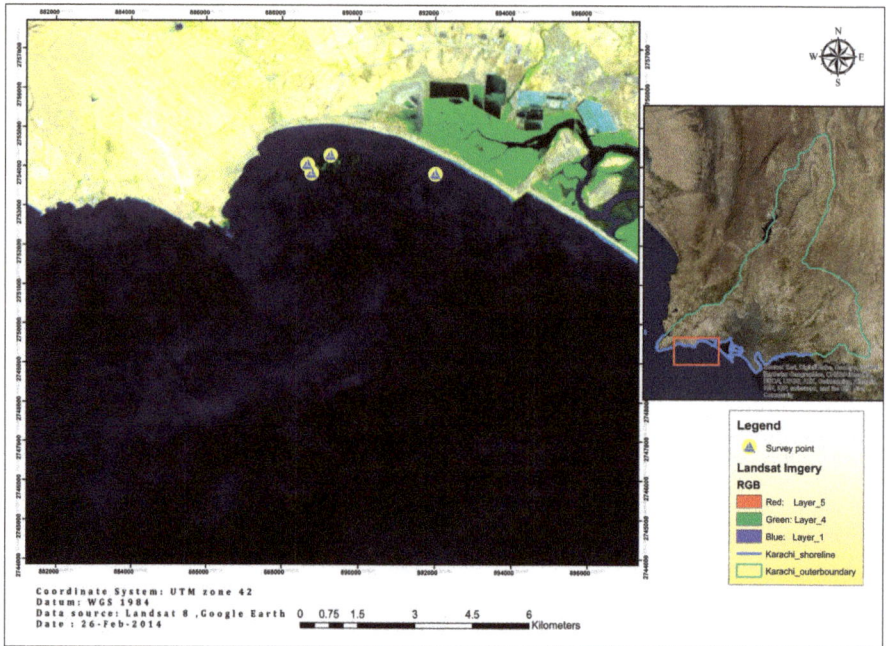

Figure 1. Study area: Hawks Bay beach along the Sindh Coast with GPS points on the seaweed patches.

Many researchers have used moderate resolution imaging spectroradiometer (MODIS), medium resolution imaging spectrometer (MERIS), and Landsat satellite data to study floating algae and seaweed indices [11–13]. MERIS 30 m data are available only for few regions of the world. MODIS has a coarser spatial resolution to monitor floating algae seaweed and, therefore, not useful in mapping small patches. In MODIS 250 m data, not every pixel is algae, so there can be mixed pixels having algae with water [13]. In this study, nine cloud-free Landsat 8 satellite images of years 2014, 2016, and 2018 in the seaweed growing months of January, February, and March were acquired and analyzed to extract seaweed patches using three different indices. Besides two commonly used bands combinations—floating algae index (FAI) and normalized difference vegetation index (NDVI)—a new seaweed enhancing index (SEI) was proposed to map seaweed patches at the study site.

2.2. Methodology

2.2.1. Pre Processing of Data

Layer stacking of all Landsat 8 bands, except the coastal/aerosol and thermal bands, was done followed by the extraction of the region under study. Digital numbers (DN) represents the pixel values of satellite images that need to be converted into reflectance values. For this purpose, top of atmospheric (TOA) reflectance was calculated using Landsat 8 operational land imager (OLI) bands from the reflectance rescaling coefficients provided in the product metadata file. Conversion of the DN of OLI data to TOA reflectance ($\rho'\lambda$), without correction for the solar angle, was performed using Equations (1) and (2) [14].

$$\rho'\lambda = MpQcal + Ap \tag{1}$$

$$\rho'\lambda = \rho'\lambda cos/(\theta SZ) = \rho'\lambda sin/(\theta SE) \tag{2}$$

where:

Mp = band-specific multiplicative from metadata;

Ap = additive rescaling factors from metadata;

$Qcal$ = quantized and calibrated standard product pixel values;

θSE = sun elevation angle;

θSZ = solar zenith angle computed by (90° − θSE);

$\rho'\lambda$ = TOA reflectance value with a correction ($\varrho\lambda$) for the sun angle was computed by equation 2 because $\rho'\lambda$ does not contain the corrected sun angle.

2.2.2. Commonly Used Band Ratios—FAI and NDVI

Floating algae on the water surface have higher reflectance in the near-infrared (NIR) than other wavelengths and thus can be easily distinguished from the surrounding clear waters. Equations 3 and 4 are used to calculate the floating algae index (FAI) [15]. Various studies have used FAI for mapping floating algae in many aquatic environments. FAI has been successfully used to detect a large bloom of floating green microalgae, *Enteromorpha prolifera*, in the open ocean near Qingdao in China under a range of atmospheric environments (clear, hazy, and sunlight conditions) [16]. FAI was found capable of discriminating between algae and water pixels. Therefore, to map floating algae in oceans, FAI is considered to be an improved index than NDVI and the enhanced vegetative index (EVI) that have limitations in detecting floating algal blooms [17]. In some research papers, FAI has also been used for detecting coastal green tides. Owing to wide recognition of FAI as an effective index for mapping floating algae, FAI was preferred to be the reference index for assessing the performance of SEI while comparing it with another generally accepted vegetation index NDVI.

$$FAI = Rrc\,NIR - Rrc\,NIR' \tag{3}$$

where:

$RrcNIR$ = baseline reflectance of NIR band.

$RrcNIR'$ can be calculated using Equation (4).

$$Rrc\,NIR' = Rrc\,(Red) + (Rrc\,(SWIR) - Rrc\,(Red)) * (\lambda NIR - \lambda Red)/(\lambda SWIR - \lambda Red) \tag{4}$$

where:

$Rrc\,(Red)$ = baseline reflectance of the red band;

$Rrc\,(SWIR)$ = baseline reflectance of the shortwave infrared (SWIR) band;

λNIR = wavelength of the NIR band;

λRed = wavelength of the red band.

For green plant remote sensing, vegetation indices are developed using the difference of the reflectance values in the NIR and red spectrum regions. The normalized difference vegetation index (NDVI) is a modest quantitative approach to measure the extent of vegetation biomass bases on these two bands as presented by Equation 5 [18]. However, the traditional vegetation indices, including NDVI, may not be very useful to study plants that are submerged or partially- submerged in water [19].

$$NDVI = (NIR - Red)/(NIR + Red) \tag{5}$$

2.2.3. Spectral Signatures and Proposed Seaweed Index

The variations of spectral signatures in reflected and absorbed electromagnetic radiation at different wavelengths help to identify specific objects. Scientist C. Hu stated that the extent of reflectance and absorption depends on the wavelength of electromagnetic radiation for any specified object. Each substrate has a different spectral signature that can be helpful to differentiate it from others, and this technique is applicable in the benthic environment [20].

Spectral characteristics of the mangroves, water, and seaweed sites in the Landsat 8 (reflectance) image were examined. The seaweed sites were identified through field surveys and recorded as GPS points. Additional GPS points were taken from a study on the submerged habitat along the Karachi Coast, which was conducted by a local marine scientist through scuba survey in February 2016 [21]. The overlay of GPS points on satellite imagery helped to capture the spectral signatures of seaweed and to develop SEI index. These signatures show meaningful peaks in NIR and SWIR bands (Bands 5 and 6, respectively) at seaweed locations differentiating water from seaweed (in Figure 2). For seaweed pixels, the high peak was observed in the NIR band (Band 5), whereas, the lowest peak was in the SWIR (Band 6) region of the electromagnetic spectrum. In NIR electromagnetic spectrum portion (700–1600 nm), macrophytes seagrasses, and seaweeds show strong reflectance since water does not fully attenuate it by generating a peak in the red shifted portion relative to those produced by the chlorophyll pigment [22]. Similar to the algorithms used in all other normalized difference indices, these two bands were used to develop a new index for seaweed, as presented in Equation 6. The new index was named the seaweed enhancing index (SEI). It is important to note that a similar trend exists for mangrove as well, though with relatively lower peaks. Therefore, it was necessary to either mask/remove mangrove area from the study area to avoid misinterpretation of mangrove pixels as seaweed or carefully examine the range of SEI to differentiate between the two substrate categories.

$$Seaweed\ Enhancing\ Index\ (SEI) = (NIR - SWIR)/(NIR + SWIR) \tag{6}$$

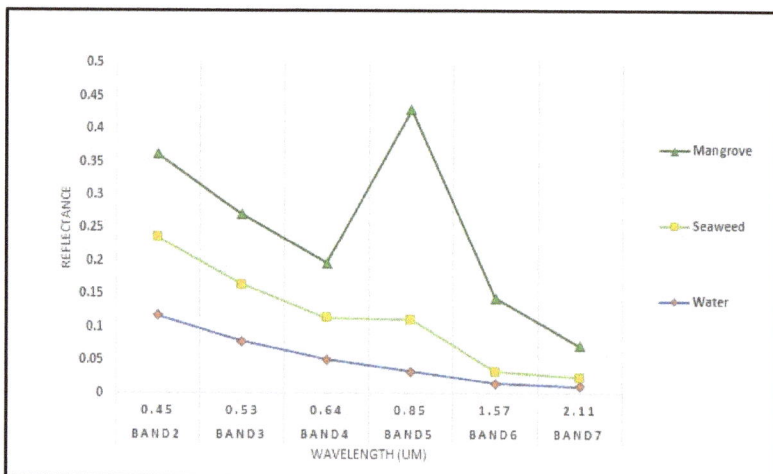

Figure 2. Spectral response of seaweed, mangrove, and water. On the x-axis Landsat 8 bands are shown and on the y-axis top of atmosphere (ToA) reflectance values are presented.

2.2.4. Extraction of Seaweed Pixel

Images of NDVI, FAI, and the newly developed index (SEI) were analyzed carefully to assign pixel value ranges for seaweed, mangrove, and water. The threshold values were set for each object class using field information. Once the seaweed pixels were defined in each index image, these were delineated as seaweed pixels. These images were converted into binary images indicating '1' as seaweed pixels and '0' as non-seaweed pixels. SEI and NDVI images were overlapped on the FAI image of the same date. Three types of pixels on each SEI image were counted, and their areas were calculated in square kilometers: (1) Seaweed pixels overlapping seaweed pixels of FAI, (2) non-seaweed pixels overlapping seaweed pixels of FAI image, and (3) seaweed pixels overlapping non-seaweed pixels of FAI. The same process was repeated for the NDVI image. Since FAI was picked as the reference index for assessing the accuracy of SEI and NDVI in extracting seaweed, more overlapping with FAI seaweed pixels was considered as an indicator of higher accuracy.

3. Results

3.1. Threshold Values for Seaweed Pixels

In the FAI image, pixel values ranged from −0.51 to 0.53, as shown in (Figure 3). After matching the seaweed sites with the pixel values, a 0.008 to 0.13 range was identified for seaweed pixels. Maximum and minimum pixels values for water were −0.51 and −0.008, respectively. The mangrove pixel value range was identified as 0.13 to 0.53 (Table 1).

Figure 3. This figure shows the floating algae index (FAI) with values ranging from −0.51 to 0.53. The red oval highlights the seaweed patches (26 February 2014).

Table 1. Floating algae index (FAI) values.

Class	Pixel Value Range
Water	−0.510 to −0.008
Seaweed	0.008 to 0.130
Mangrove	0.130 to 0.530

The normalized difference vegetation index was applied to the same image, and pixel values for seaweed, water, and mangroves classes were identified. In the NDVI image, pixel values ranged from −0.145 to 0.372 (Figure 4). The maximum value of seaweed pixel was 0.121, and the minimum value was −0.044. Maximum and minimum water pixel values were −0.145 to −0.044, and for mangrove, this range was 0.159 to 0.372 (Table 2).

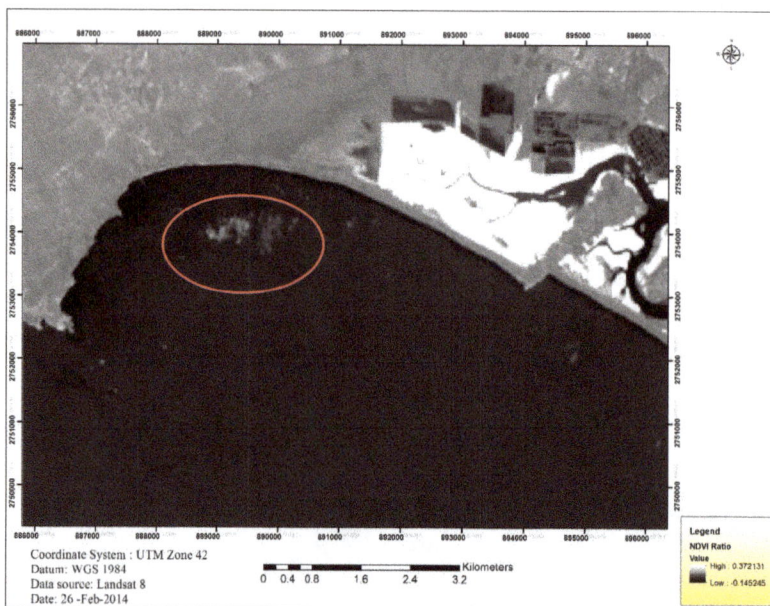

Figure 4. This figure shows NDVI with values ranging from −0.145 to 0.372. The red oval highlights the seaweed patches (26 February 2014).

Table 2. Normalized difference vegetation index (NDVI) values.

Class	Pixel Value Range
Water	−0.145 to −0.044
Seaweed	−0.044 to 0.121
Mangrove	0.159 to 0.372

Seaweed pixel values for SEI had a range from 0.08 to 0.24, as shown in (Figure 5). Maximum and minimum values of the water pixels, respectively, were −0.08 and −0.85, whereas, the mangrove pixel values ranged from 0.24 to 0.307, as presented in Table 3.

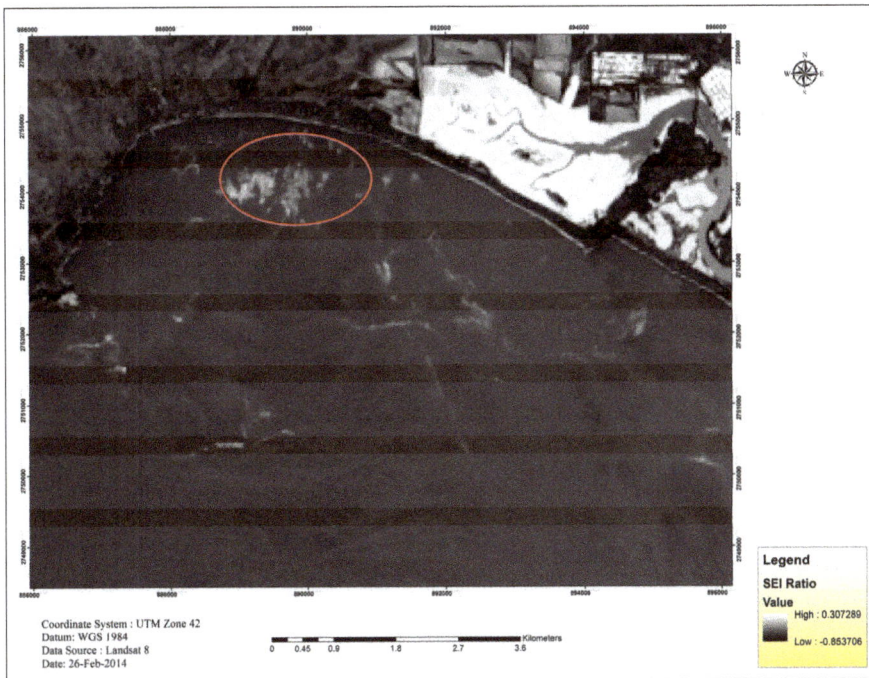

Figure 5. This figure shows the seaweed enhancing index (SEI) with values ranging from −0.85 to 0.307. The red oval highlights the seaweed patches (26 February 2014).

Table 3. SEI index values.

Class	Pixel Value Range
Water	−0.85 to 0.08
Seaweed	0.08 to 0.24
Mangrove	0.24 to 0.307

3.2. Seaweed Area Estimation

Three even years, 2014, 2016, and 2018 with the growing seaweed months of January, February, and March, were selected at two year intervals as the study period. A total of nine images were utilized to assess the performance of SEI at a more extended period to avoid any temporary site-specific anomalies. For all temporally separated images, the spectral signatures of seaweed were found similar. During analysis, it was observed that seaweed patches in all images were at exactly 2014 collected GPS points indicating them as permanent seaweed sites. Dense seaweed was found in January and February and was sparse in March, which may be due to its nearness with the seaweed ending season. A summary of the estimated areas is presented in Table 4.

Table 4. Seaweed area estimation (years 2014, 2016, and 2018).

Landsat 8 Image Date and Month	Indices	Area Estimation (km^2)
25 January 2014	NDVI	1.78
	FAI	1.91
	SEI	1.93
19 February 2014	NDVI	0.98
	FAI	1.97
	SEI	2.1
7 March 2014	NDVI	0.594
	FAI	0.75
	SEI	0.79
24 January 2016	NDVI	0.59
	FAI	0.63
	SEI	0.65
9 February 2016	NDVI	0.7
	FAI	1.1
	SEI	1.54
3 March 2016	NDVI	0.908
	FAI	1.8
	SEI	1.95
21 February 2018	NDVI	0.30
	FAI	0.34
	SEI	0.42
9 March 2018	NDVI	0.190
	FAI	0.32
	SEI	0.43

The image of 24 January 2016, was slightly cloudy and NDVI showed some mixed pixels and enhanced some non-seaweed areas. In 2018, only the February and March months had cloud-free images, and therefore, were selected for this study. During analysis, it was noted that in 2018 seaweed patches were fewer in quantity as compared to other years.

SEI and FAI greatly enhanced seaweed patches, whereas, NDVI estimated area was the lowest among all. During validation of all three indices with GPS points, it was observed that FAI and NDVI did not enhance attached seaweed, although SEI enhanced attached patches, it also mapped some rocky areas as seaweed. The radar graph, as shown in Figure 6, indicated the area extracted in all indices. The outer part illustrates the timeline, whereas, the inner loops from 0 to 2 show area values.

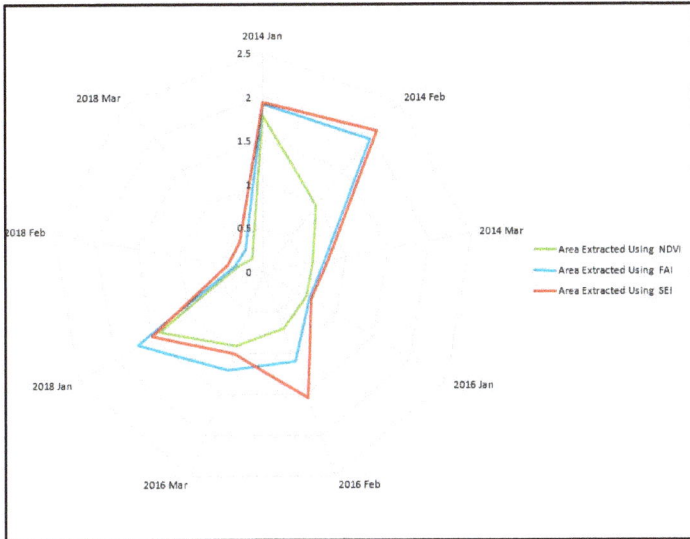

Figure 6. Graph showing estimated areas in all indices.

3.3. Validation of Seaweed Enhancing Index (SEI)

For validation purposes, binary images of all indices of the same dates were overlaid. The accuracy of NDVI and SEI was assessed by matching their spatial extent of seaweed cover with FAI enhanced seaweed area (Figure 7). The raster calculator in ArcMap was used to create new images combining FAI separately with NDVI and SEI. The new images had three distinct area classes: (1) Seaweed in both FAI and SEI (NDVI), (2) seaweed in FAI but not in SEI (NDVI), and (3) seaweed in SEI (NDVI) but not in FAI. The remaining pixels in these images belonged to a non-seaweed area in FAI and SEI (NDVI). The pixels overlapping in an index and FAI as seaweed pixels indicated the accuracy of that index. A summary of areas in three classes mentioned here for both SEI and NDVI is given in Table 5. This information will tell the accuracy of SEI and NDVI conforming to FAI, which was being used as the reference index in the study.

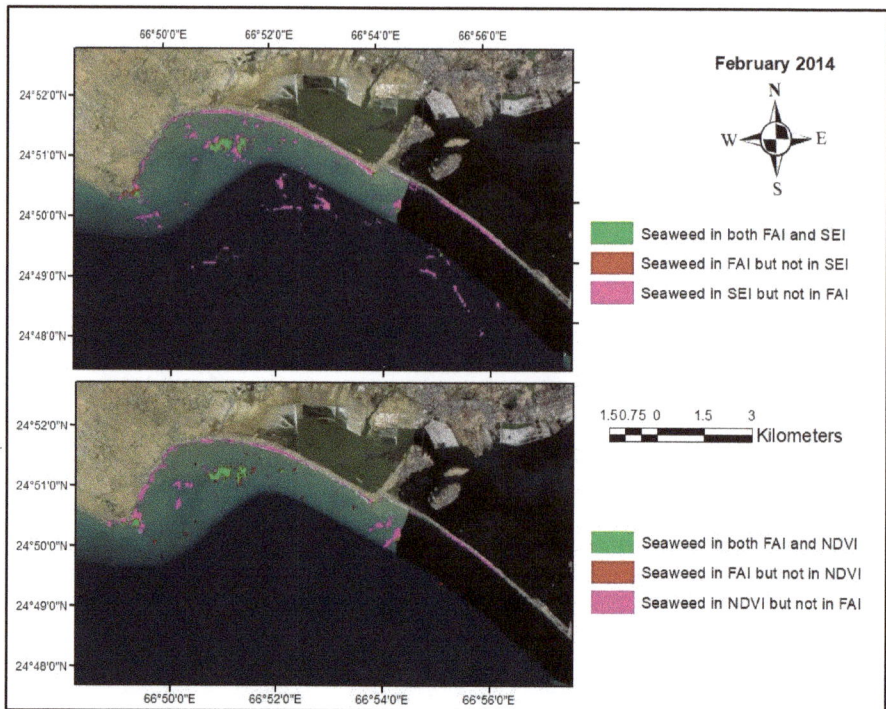

Figure 7. This figure compares seaweed extraction of SEI and NDVI with FAI.

Table 5. Areas in km^2.

	For NDVI	For SEI		For NDVI	For SEI
	Jan 2014			March 2016	
A	1.3932	1.2294	A	0.387	0.4077
B	0.2583	0.4212	B	0.2475	0.2268
C	0.0027	0.6498	C	0.0009	0.7659
	Feb 2014			Jan 2018	
A	0.3501	0.4122	A	2.7045	4.0689
B	0.1125	0.0504	B	1.4148	0.0504
C	0.7821	1.5129	C	0.4122	5.1174
	March 2014			Feb 2018	
A	0.4743	0.3942	A	0.522	0.1161
B	0.0585	0.1386	B	0.0108	0.4167
C	0.666	0.7821	C	0.0594	0.3519
	Jan 2016			March 2018	
A	0.2421	0.081	A	0.5076	0.0765
B	0	0.1611	B	0.0252	0.4563
C	2.8197	0.1188	C	0.0081	0.1188
	Feb 2016				
A	0.4374	0.612	Seaweed in both FAI and SEI/NDVI	A	
B	0.2232	0.0486	Seaweed in FAI but not in SEI/NDVI	B	
C	0.1593	0.1602	Seaweed in SEI/NDVI but not in FAI	C	

4. Discussion

The newly developed seaweed enhancing index (SEI), enhanced larger areas of seaweed resources as compared to NDVI and FAI, as shown in (Figure 5). The results showed that NDVI extracted fewer seaweed areas as compared to FAI and SEI. In NDVI, big patches of seaweed were enhanced, but it failed to enhance the attached seaweed (Figure 8). Besides NDVI, FAI also could not map attached seaweeds on rocky areas very well. However, in SEI seaweeds attached on rocks were enhanced, but also got mixed with some non-seaweed rocky pixels.

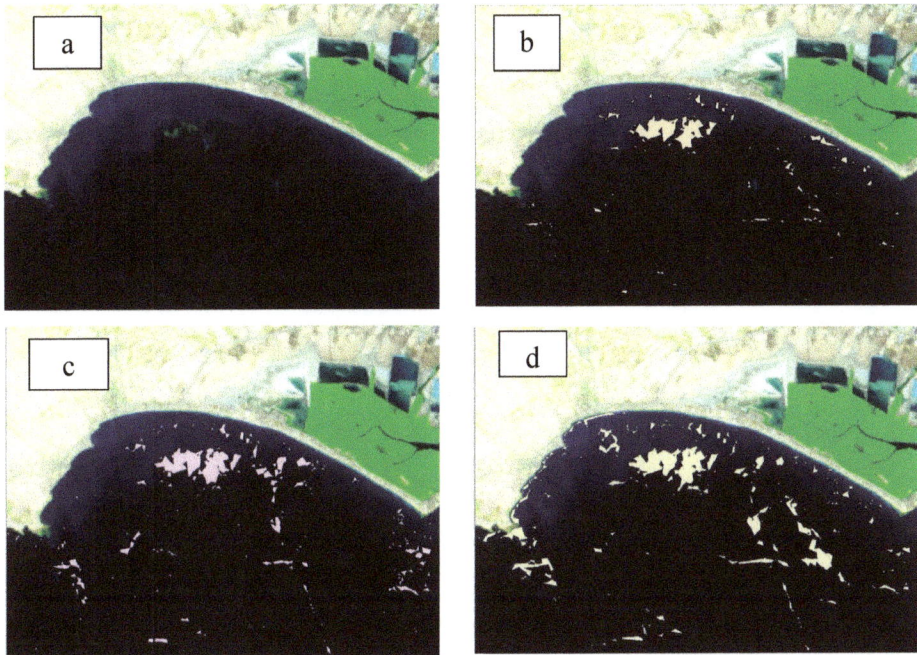

Figure 8. Seaweed extracted class overlay on Landsat images: (**a**) False color composite image, (**b**) NDVI, (**c**) FAI, and (**d**) SEI.

Seaweed areas enhanced by SEI and NDVI overlapping FAI seaweed area as the percentages of total FAI seaweed area (A/(A + B) × 100) are presented in Table 6. The SEI images of January 2016, February 2018, and March 2018 enhanced less than 50 percent of the corresponding FAI total seaweed areas. However, on these dates, the NDVI performed very well, matching more than 95 percent of FAI seaweed coverage. Except for these three years, the performance of SEI in the remaining six images was either similar to NDVI or even better than NDVI. On January 2018, SEI enhanced 99 percent of FAI seaweed cover. Overall, the seaweed area not covered by FAI was greater in SEI than NDVI in almost all images, which needs to be further explored in future studies by collecting extensive field information to validate SEI mapped additional area beyond the extent of FAI seaweed cover. Based on these results, in the majority of the satellite temporal images selected for this study, the performance of the newly proposed index—SEI—was found either better than or similar to NDVI. Except one, almost in all cases, the index area as seaweed not covered by FAI was greater in SEI than NDVI. These locations need to be validated through field collected data where SEI has mapped seaweed, but FAI has not.

Table 6. Comparing NDVI and SEI with FAI.

Image Date	Jan-14	Jan-16	Jan-18	Feb-14	Feb-16	Feb-18	Mar-14	Mar-16	Mar-18
Total seaweed area in FAI (km^2)	1.6515	0.2421	4.1193	0.4626	0.6606	0.5328	0.5328	0.6345	0.5328
NDVI (%)	84.36	100.00	65.65	75.68	66.21	97.97	89.02	60.99	95.27
SEI (%)	74.48	33.46	98.78	89.11	92.64	21.79	73.99	64.26	14.36

5. Conclusions

In this study, three indices—floating algal index (FAI), normalized difference vegetative index (NDVI), and seaweed enhancing index (SEI)—were applied on Landsat 8 temporal images to extract seaweed patches along the Karachi Coast. Analyzing satellite mapped data, the following trends were observed:

- All three indices enhanced seaweed at the verified field collected GPS locations in all temporal images, which employed that the GPS points were at the permanent (seasonal) seaweed sites.
- Area estimation of seaweed resources of three indices showed variations. Overall, SEI extracted seaweed area was more than both NDVI and FAI, which was probably due to SEI capability of enhancing rocked attached patches and overestimation of rocked area. Another reason may be the possibility of SEI enhancing the submerged aquatic vegetation, which has low SWIR reflectance values [23]. However, this needs to be investigated, and if this reasoning happens to be correct, then it means that SEI works better in enhancing partially submerged seaweed patches, which other indices fail to do.
- NDVI and FAI failed to enhance rock attached seaweed pixels.
- The performance of SEI was found either better than or similar to NDVI based on percent seaweed area of FAI overlapped by the index.

Seaweed assessment is valuable for many stakeholders, including the fisherman community, policymakers, and food and cosmetics industries. These studies can be beneficial to support coastal resources. Remote sensing techniques have been proved as a valuable tool for monitoring and mapping marine resources. In this study, a new seaweed enhancing index was introduced that had the potential to be used for seaweed mapping. This study also demanded some inquiries regarding SEI capabilities in mapping partially submerged patches. A more detailed study is needed over a longer period to validate this notion. If this hypothesis happens to be true, it will open the possibility of detecting seaweed locations more precisely with SEI.

Author Contributions: Conceptualization, M.D.S. and M.A.; methodology, M.D.S. and A.Z.Z.; software, NA; validation, M.D.S. and A.Z.Z.; formal analysis, M.D.S. ; investigation, M.D.S. and A.Z.Z.; resources, A.Z.Z.; data curation, M.D.S. and M.A.; writing—original draft preparation, M.D.S.; writing—A.Z.Z.; visualization, M.D.S.; supervision, A.Z.Z.; project administration, M.D.S. and M.A.; funding acquisition, A.Z.Z., M.D.S.

Funding: This research was partially funded by Mangroves for the Future (MFF) of IUCN, grant number, NA. The APC was waived.

Acknowledgments: We, the authors, are grateful to the Centre of Excellence in Marine Biology, the University of Karachi for arranging collaborative field survey. We would also like to knowledge Rafey Ali, a research specialist at The Aga Khan University, for his valuable assistance in this research.

Conflicts of Interest: The authors declare no conflict of interest. The funders had no role in the design of the study; in the collection, analyses, or interpretation of data; in the writing of the manuscript, or in the decision to publish the results.

References

1. Magg, W. *Seaweed—Types of Seaweed*; Te Ara-the Encyclopedia of New Zealand: Wellington, New Zealand, 2006.
2. Haq, T.; Khan, F.A.; Begum, R.; Munshi, A.B. Bioconversion of drifted seaweed biomass into organic compost collected from the Karachi coast. *Pak. J. Bot.* **2011**, *43*, 3049–3051.
3. Choi, C.G.; Takeuchi, Y.; Terawaki, T.; Serisawa, Y.; Ohno, M.; Sohn, C.H. Ecology of seaweed beds on two types of an artificial reef. *J. Appl. Phycol.* **2002**, *14*, 343–349. [CrossRef]
4. Okuda, K. The Coastal Environment and Seaweed-Bed Ecology in Japan. *Kuroshio Sci.* **2008**, *2*, 15–20.
5. Valderrama, D. Social and Economic Dimensions of Seaweed Farming: A Global Review. In Proceedings of the 16th International institute of fisheries economics and trade (IIFET), Dar es Salaam, Tanzania, 16–20 July 2012.
6. Capuzzo, E.; McKie, T. *Seaweed in the UK and Abroad—Status, Products, Limitations, Gaps, and Cefas Role*; Centre for Environment, Fisheries & Aquaculture Science (Cefas): Suffolk, UK, 2016.

7. Samee, H.; Li, Z.X.; Lin, H.; Khalid, J.; Guo, Y.C. Anti-allergic effects of ethanol extract from brown seaweeds. *J. Zhejiang Univ. Sci. B* **2009**, *10*, 147–153. [CrossRef] [PubMed]
8. Pickrill, R.A.; Kostylev, V.E. Habitat Mapping and National Seafloor Mapping Strategies in Canada. *Geol. Assoc. Canada* **2007**, *47*, 449–462.
9. Eakin, C.M.; Nim, C.J.; Brainard, R.E.; Aubrecht, C.; Elvidge, C.; Gledhill, D.K.; Muller-Karger, F.; Mumby, P.J.; Skirving, W.J.; Strong, A.E.; et al. Monitoring coral reefs from space. *Oceanography* **2010**, *23*, 118–133. [CrossRef]
10. Hyun, J.C.; Deepak, M.; John, W. Remote Sensing of Submerged Aquatic Vegetation. In *Remote Sensing—Applications*; IntechOpen: London, UK, 2012.
11. He, M.X.; Liu, J.; Yu, F.; Li, D.; Hu, C. Monitoring green tides in Chinese marginal seas. In *Handbook of Satellite Remote Sensing Image Interpretation: Applications for Marine Living resources conservation and management*; Morales, J., Stuart, V., Platt, T., Sathyendranath, S., Eds.; EU PRESPO and IOCCG: Dartmouth, NS, Canada, 2011; pp. 111–124.
12. Gower, J.; Hu, C.; Borstad, G.; King, S. Ocean color satellites show extensive lines of floating Sargassum in the Gulf of Mexico. *IEEE Trans. Geosci. Remote Sens.* **2006**, *44*, 3619–3625. [CrossRef]
13. Gower, J.; King, S.; Goncalves, P. Global monitoring of plankton blooms using MERIS MCI. *Int. J. Remote Sens.* **2008**, *29*, 6209–6216. [CrossRef]
14. Ko, B.; Kim, H.; Nam, J. Classification of potential water bodies using Landsat 8 OLI and a combination of two boosted random forest classifiers. *Sensors* **2015**, *15*, 13763–13777. [CrossRef] [PubMed]
15. Hu, C. A novel ocean color index to detect floating algae in the global oceans. *Remote Sens. Environ.* **2009**, *113*, 2118–2129. [CrossRef]
16. Hu, C.; He, M. Origin and offshore extent of floating algae in Olympic sailing area. *Eos Trans. Am. Geophys. Union* **2008**, *89*, 302. [CrossRef]
17. El-Alem, A.; Chokmani, K.; Laurion, I.; El-Adlouni, S.E. Comparative analysis of four models to estimate chlorophyll-a concentration in case-2 waters using MODerate resolution imaging spectroradiometer (MODIS) imagery. *Remote Sens.* **2012**, *4*, 2373–2400. [CrossRef]
18. Campbell, J.B.; Wynne, R.H. *Introduction to Remote Sensing*; Guilford Press: New York, NY, USA, 2011.
19. Cho, H.J.; Lu, D. A water-depth correction algorithm for submerged vegetation spectra. *Remote Sens. Lett.* **2010**, *1*, 29–35. [CrossRef]
20. Lubin, D.; Li, W.; Dustan, P.; Mazel, C.H.; Stamnes, K. Spectral signatures of coral reefs: Features from space. *Remote Sens. Environ.* **2001**, *75*, 127–137. [CrossRef]
21. Ali, A.; Malik, S.; Zaidi, A.Z.; Ahmad, N.; Shafique, S.; Aftab, M.N.; Aisha, K. Standing Stock Of Seaweeds In Submerged Habitats Along The Karachi Coast, Pakistan: An Alternative Source Of Livelihood For Coastal Communities. *Pak. J. Bot.* **2019**, *51*, 5. [CrossRef]
22. Dierssen, H.M.; Zimmerman, R.C.; Bissett, P.J. The Red Edge: Exploring high near-infrared reflectance of phytoplankton and submerged macrophytes and implications for aquatic remote sensing. In *AGU Spring Meeting Abstracts*; American Geophysical Union (AGU): Washington, DC, USA, 2007.
23. Hestir, E.L. *Trends in Estuarine Water Quality and Submerged Aquatic Vegetation Invasion*; University of California: Davis, CA, USA, 2010.

remote sensing

MDPI

Article

Rapid Invasion of *Spartina alterniflora* in the Coastal Zone of Mainland China: New Observations from Landsat OLI Images

Mingyue Liu [1,2], Dehua Mao [1,*], Zongming Wang [1,*], Lin Li [3], Weidong Man [1,2], Mingming Jia [1], Chunying Ren [1] and Yuanzhi Zhang [4,5,*]

[1] Key Laboratory of Wetland Ecology and Environment, Northeast Institute of Geography and Agroecology, Chinese Academy of Sciences, Changchun 130102, China; liumy917@ncst.edu.cn (M.L.); manwd@ncst.edu.cn (W.M.); jiamingming@iga.ac.cn (M.J.); renchy@iga.ac.cn (C.R.)
[2] College of Mining Engineering, North China University of Science and Technology, Tangshan 063210, China
[3] Department of Earth Sciences, Indiana University-Purdue University, Indianapolis, IN 46202, USA; ll3@iupui.edu
[4] Chinese University of Hong Kong, Center for Housing Innovations, Shatin, New Territories, Hong Kong, China
[5] Chinese Academy of Sciences, Key Lab of Lunar Science and Deep-exploration, National Astronomical Observatories, Beijing 100101, China
* Correspondence: maodehua@iga.ac.cn (D.M.); zongmingwang@iga.ac.cn (Z.W.); yuanzhizhang@cuhk.edu.hk (Y.Z.); Tel.: +86-431-85542254 (D.M.)

Received: 1 November 2018; Accepted: 28 November 2018; Published: 1 December 2018

Abstract: Plant invasion imposes significant threats to biodiversity and ecosystem function. Thus, monitoring the spatial pattern of invasive plants is vital for effective ecosystem management. *Spartina alterniflora* (*S. alterniflora*) has been one of the most prevalent invasive plants along the China coast, and its spread has had severe ecological consequences. Here, we provide new observation from Landsat operational land imager (OLI) images. Specifically, 43 Landsat-8 OLI images from 2014 to 2016, a combination of object-based image analysis (OBIA) and support vector machine (SVM) methods, and field surveys covering the whole coast were used to construct an up-to-date dataset for 2015 and investigate the spatial variability of *S. alterniflora* in the coastal zone of mainland China. The classification results achieved good estimation, with a kappa coefficient of 0.86 and 96% overall accuracy. Our results revealed that there was approximately 545.80 km^2 of *S. alterniflora* distributed in the coastal zone of mainland China in 2015, from Hebei to Guangxi provinces. Nearly 92% of the total area of *S. alterniflora* was distributed within four provinces: Jiangsu, Shanghai, Zhejiang, and Fujian. Seven national nature reserves invaded by *S. alterniflora* encompassed approximately one-third (174.35 km^2) of the total area of *S. alterniflora* over mainland China. The Yancheng National Nature Reserve exhibited the largest area of *S. alterniflora* (115.62 km^2) among the reserves. Given the rapid and extensive expansion of *S. alterniflora* in the 40 years since its introduction and its various ecological effects, geospatially varied responding decisions are needed to promote sustainable coastal ecosystems.

Keywords: invasive plants; *Spartina alterniflora*; CAS *S. alterniflora*; object-based image analysis; Landsat OLI

---

1. Introduction

Plant invasion, as an important type of biological invasion, has emerged as a serious ecological issue, which threatens native species and affects the structure and function of ecosystems [1–4]. In coastal zones, widespread invasive plants have strong impacts on biogeochemical cycles and thus have severe environmental consequences [5–7]. Thus, particular attention to the invasive plants in coastal area is necessary to ensure ecological security and maintain sustainable ecosystems.

Spartina alterniflora (*S. alterniflora*) has been categorized as one of the most serious invasive plants by the State Environmental Protection Administration of China. The invasion of this exotic species has had vast negative consequences, including threatening native wetland plants and waterfowls, and imposing negative effects on fishing, water transportation, mariculture activities, and tourism development [8–10]. *S. alterniflora* was first introduced from the Atlantic coast of the United States (U.S.) to China in 1979 for the purpose of tidal land reclamation, seashore stabilization, and saline soil amelioration [10–12]. Previous studies have documented that the area of invasive *S. alterniflora* in coastal China exceeds that of mangroves [12–14]. China has been the largest country invaded by exotic *S. alterniflora*. Although *S. alterniflora* has great potential for carbon sequestration and biofuel due to its high productivity and strong adaptability [10,15], the sustainable management of China's coastal zone requires the acquisition of additional quantitative data to effectively respond to the expansion of *S. alterniflora* and its consequences. In particular, the up-to-date spatial information of *S. alterniflora* at the national scale, 40 years since its introduction, is necessary for coastal ecosystem conservation and economic development.

Remote sensing has been identified as an effective tool for detecting invasive plants [16–19]. The selection of a suitable data source and a classification method is commonly case-specific and largely depends on the target plant and research goals [20,21]. For example, synoptic aerial photographs from 1945 to 2000 were used to characterize the spatiotemporal patterns of *S. alterniflora* in Willapa Bay in the U.S. [22]. High/ultra-high spatial resolution images, such as SPOT 6 and unmanned aerial vehicle (UAV) images, were used to obtain detailed distributions of *S. alterniflora* in China's Yueqing Bay [11] and Beihai city [23], respectively. Freely available Google Earth images with high spatial resolution were employed to identify *S. alterniflora* invasion to mangroves in Zhangjiang Estuary [24]. Generally, satellite images with moderate spatial resolution, such as Landsat and China–Brazil Earth Resource Satellite (CBERS) images, are suitable data sources for mapping the distribution of invasive *S. alterniflora* at large scales [10,12,25]. Compared to the currently accessible data sources, the newly launched Landsat 8, which carries the operational land imager (OLI) sensor, provides more easy-to-access, high-quality images due to its intensive image acquisition capability and improved duty cycle [26]. Thus, Landsat 8 allows the possibility of mapping the distribution of *S. alterniflora* along the 18,000 km of China's coast [27] for a specific time period. Moreover, an increasing number of studies have adopted object-based image analysis (OBIA) to identify *S. alterniflora* [23,24,28,29], and the OBIA and support vector machine (SVM) have been proven promising for mapping the invasion of *S. alterniflora* [11]. These data source and image classification method developments could greatly contribute to updating the invasion information of *S. alterniflora*.

In mainland China, most of the previous studies of *S. alterniflora* invasion have focused on local regions, and thus failed to update the spatial distribution *S. alterniflora* in a timely manner, even though this information is critical for supporting land management, protecting important habitats of endangered species, and ensuring ecological security in response to global change. The resulting information deficiency has limited decision-making regarding the sustainable ecosystem management of coastal wetlands and the socioeconomic development of coastal cities. To address this deficiency, this study aimed to provide a new observation using Landsat OLI images and the integration of OBIA and SVM. Specifically, this study mapped the up-to-date distribution of *S. alterniflora* at the national scale, and documented the spatial variation in invasion status. The finding in this study could provide important quantified areal data for the ecological studies of *S. alterniflora*, and is also a baseline dataset for documenting the spatiotemporal dynamics of *S. alterniflora* invasion.

2. Materials and Methods

2.1. Study Area

Considering the ecological niche of *S. alterniflora* and the common definition of coastal zone in China, we defined the contiguous region extending from the landward 10-km buffer line of the coastline over mainland China to the first continuous contour of 15-m water depth, which was derived from the global relied model, as the study area. The study area is located in the coastal zone of mainland China (Figure 1), which spans 10 provinces (Liaoning, Hebei, Tianjin, Shandong, Jiangsu, Shanghai, Zhejiang, Fujian, Guangdong, and Guangxi). This zone covers the warm temperate zone, subtropical zone, and tropical zone from north to south. Wetland is the dominant ecosystem type, while the common wetland plants include *Phragmites australis, S. alterniflora, Suaeda salsa, Tamarix chinensis, Scirpus mariquete, Cyperus malaccensis*, and mangrove forests. *S. alterniflora* grows widely in the intertidal zone, and tends to spread parallel to and continuous along shorelines. This alien species can colonize a variety of substrates, ranging from sand and silt to loose cobbles, clay, and gravel.

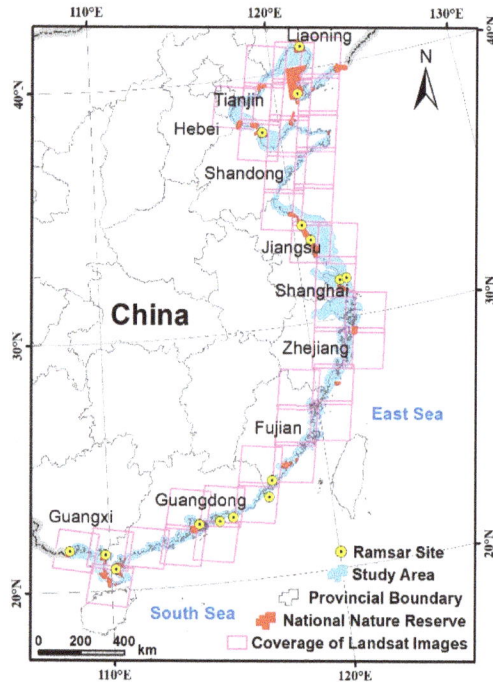

Figure 1. Location of the study area and the distribution of *Ramsar* sites and national nature reserves.

2.2. Data and Preprocessing

2.2.1. Landsat Imagery

In this study, 43 scenes of Landsat 8 OLI images from 2014 to 2016 were selected to delineate *S. alterniflora* in the coastal zone of mainland China. These images were downloaded from the United States Geological Survey (USGS, https://glovis.usgs.gov/). An optimal acquisition time is important to accurately discriminate *S. alterniflora* from other salt marsh plants. For this study, OLI images were selected by considering the phenological divergence of local species. Generally, vegetation in the peak growing period may show significant spectral similarity. Therefore, OLI images acquired in the spring and autumn are generally preferred to use for *S. alterniflora* identification in the northern provinces, whereas those in early spring and winter are used for identification in the southern provinces [10,30,31]. Multiple scenes of images were also used to enhance the separation of *S. alterniflora* from other species by considering the phenological stages and tidal level. A total of 33 scenes of Landsat images could cover the whole study area. We used an additional 10 images to support the image classification. For example, the images in spring or autumn could be used for separating the *S. alterniflora* from mangrove, because the mangrove is an evergreen species.

2.2.2. DEM and ETOPO 1 Data

Digital elevation model (DEM) tiles derived from the Advanced Spaceborne Thermal Emission and Reflection Radiometer Global DEM version2 (ASTER GDEM v2) at approximately 30-m resolution were downloaded from the USGS site. ETOPO1 is a one arc-minute global relief model of the Earth's surface that integrates land topography and ocean bathymetry, which was obtained from the National Oceanic and Atmospheric Administration (NOAA, http://dx.doi.org/10.7289/V5C8276M).

2.2.3. National Nature Reserves

For protecting coastal wetland ecosystems and endangered animals, to date, 15 wetland sites with international importance (Ramsar sites) and 32 national nature reserves (NNRs) have been established in the coastal zone of mainland China (Figure 1). In this study, the NNR boundary dataset was obtained to document the invasive status in NNRs and compare the difference among the different functional zones (core zone, buffer zone, and experimental zone) of NNRs. Based on the administrative regulations of national nature reserves in China, the experimental zone of a national reserve can develop activities of breeding rare and endangered animal or plant species, teaching practice, and tourism. The buffer zone could have only limited scientific research activities, while the core zone should not have any human activity.

2.2.4. Field Surveys

Field surveys were conducted between September and November from 2014 to 2016 along the shoreline of mainland China (Figure 2) to collect ground truth points. Some sites were investigated by unmanned aerial vehicle due to road inaccessibility. A total of 11085 of land cover points were recorded using a hand-held geographic positioning system, of which 1716 were of *S. alterniflora*. We randomly collected 70% of the ground truth points as training samples, and another 30% as validation samples. Specifically, 1201 *S. alterniflora* and 6558 other land cover points were randomly selected as training samples, and 515 *S. alterniflora* and 2811 *non-S. alterniflora* points were used as validation samples in the image classification.

Due to road inaccessibility and bad weather when we carried out the field investigation in Guangdong Province, limited field truth samples were obtained in this province. Previous studies have revealed that there were only a few areas of *S. alterniflora* in Guangdong Province. Therefore, we collected 34 samples of *S. alterniflora* from the high-resolution images of Google Earth and other published papers for the training process of object identification in five scenes of Landsat images.

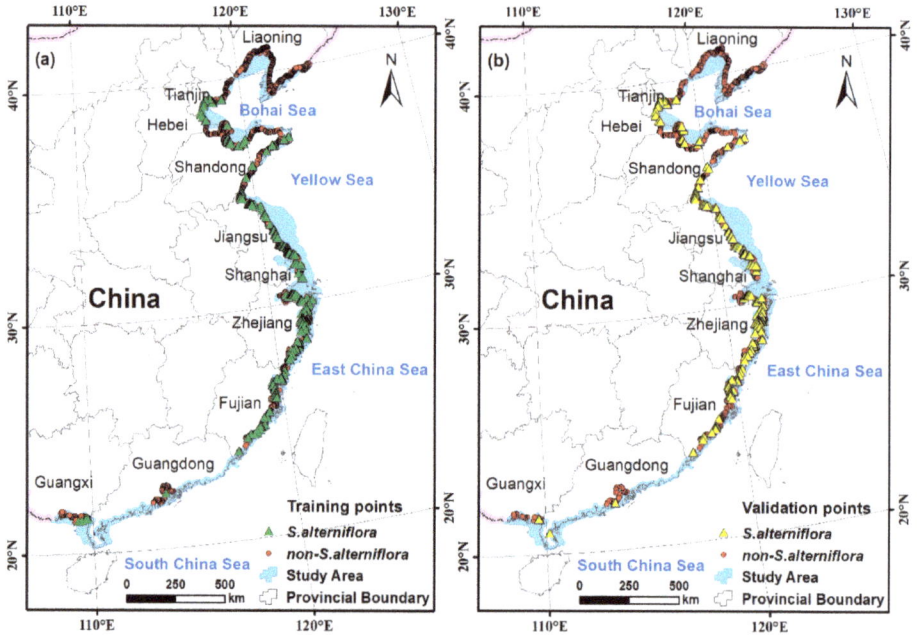

Figure 2. Distribution of field survey observations: (**a**) training samples; (**b**) validation samples.

2.2.5. Data Preprocessing

In this study, all of the images were processed for atmospheric correction using the Fast Line-of-sight Atmospheric Analysis of Hypercubes (FLAASH) model and georectified to 1:100,000 topographic maps using ground control points (GCPs) in the ENVI 5.0 image processing software package. To improve classification accuracy, the OLI panchromatic band with a spatial resolution of 15 m was used together with seven multispectral bands with a spatial resolution of 30 m in the process of image segmentation. All of the images, reference data, and field survey shapefiles were projected to the Albers equal area conic projection with the datum WGS 84 coordinate system. Before we performed the image segmentation, all of the images were clipped using the boundary of study area.

2.3. Extracting the Distribution of S. alterniflora

In this study, we combined OBIA and SVM methods to extract the *S. alterniflora*. In the process of OBIA, textural, geometric, and contextual features at the object level, as well as spectral information, were combined to provide a rich pool of candidate variables for classification [32,33]. SVM is a supervised non-parametric statistical learning technique that is suitable for performing non-linear, high-dimensional space classifications of remote sensing imagery [34,35]. These two functions built in the eCognition Developer 9 software were used to extract *S. alterniflora*. The input image layers were composed of the panchromatic and multispectral bands of the OLI image, DEM, and ETOPO1 data. In addition, the shapefile of the training samples was imported as a thematic layer to identify object samples for training the SVM classifier. The Fuzzy-based Segmentation Parameter (FbSP) optimizer was used to determine the optimal parameters for multi-resolution segmentation instead of employing the traditional trial and error method. Figure 3 is the flowchart for extracting *S. alterniflora*.

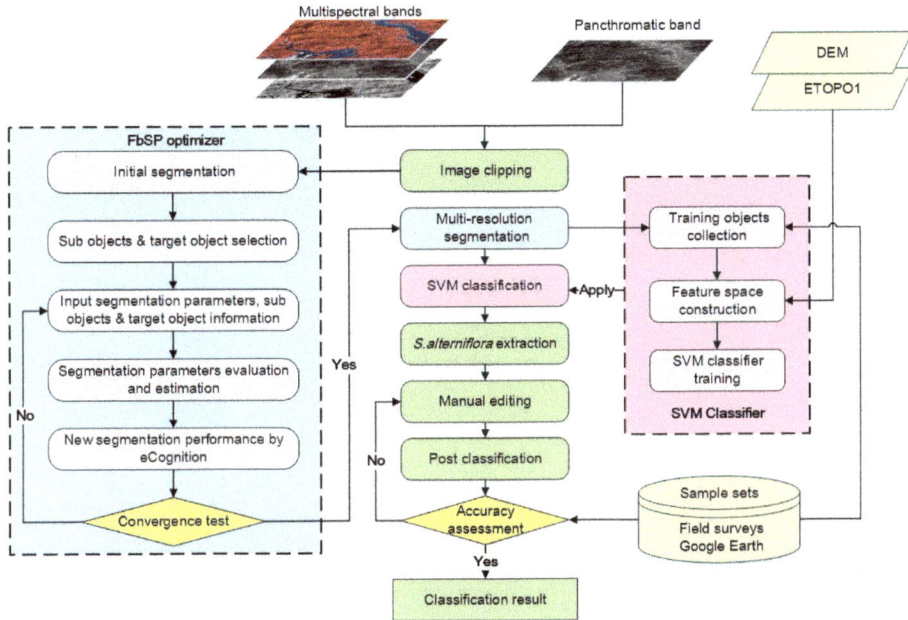

Figure 3. Flowchart of the processing scheme for identifying *S. alterniflora*.

2.3.1. Multiscale Segmentation

Segmentation is the first key step in the OBIA process, and its outputs provide the foundation for subsequent classification that directly influences the classification accuracy. The FbSP optimizer, a commonly used multi-resolution segmentation algorithm [36,37], was applied to objectively determine the optimal segmentation parameters (scale, shape/color, and smoothness/compactness). The FbSP optimizer was developed based on the idea of discrepancy evaluation to control the merging process of sub-objects and work through a supervised training process and fuzzy logic analysis [36]. Specifically, an initial segmentation of input images, which achieves an excessive segmentation result, was performed. The default eCognition settings for the shape and compactness parameters, and a small value of scale parameter, were normally used to generate sub-objects, which are smaller than the target object. Sub-objects were then selected from the initial segmentation result as training objects, and their values of related features, including texture, stability, brightness, and area were collected. Further, the training objects were merged, and the feature values of merged objects were collected. Both the feature values of sub-objects and merged objects were imported into the FbSP optimizer to generate new segmentation parameters. The parameters provided by the FbSP optimizer were used to segment images again using eCognition software. Such a training process was iteratively performed to reach a convergence between segmentation and the target object until they match each other. The optimal segmentation parameters are thus obtained. This training process was performed for each scene of image. Figure 4 shows an example of the segmentation process based on the FbSP optimizer.

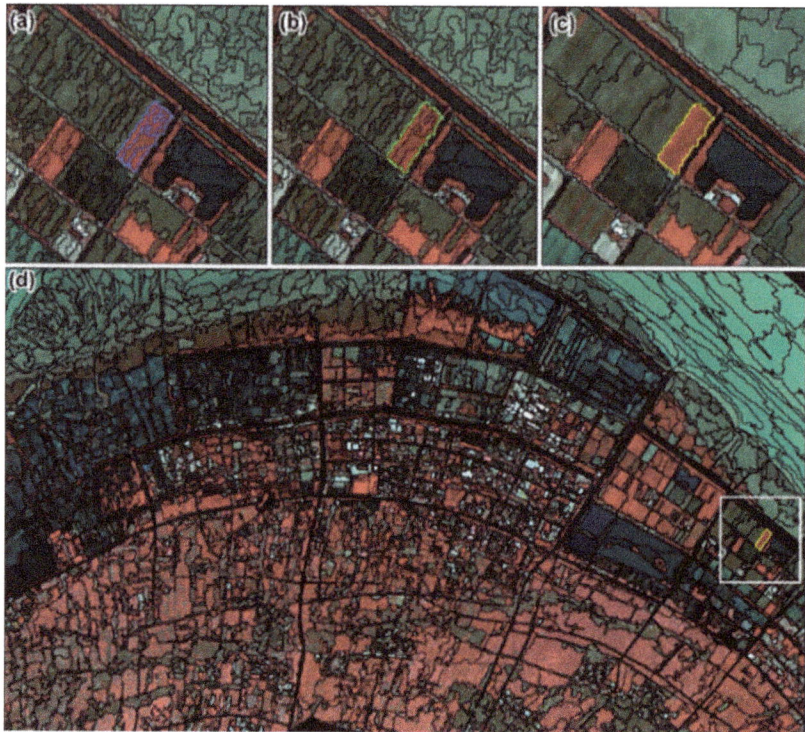

Figure 4. An example of the segmentation process and final segmentation result using the Fuzzy-based Segmentation Parameter (FbSP) optimizer (image path/row: 118/39). (**a**) Initial segmentation and sub-objects of vegetation (blue); (**b**) target object (green) formed by the sub-objects (the blue in (**a**)) for training the FbSP optimizer; (**c**) the object (yellow) resulting from the second segmentation iteration generated by using the parameters estimated by the FbSP optimizer that achieved convergence with the target object (the green in (**b**)); (**d**) final segmentation result yielded by using the parameters estimated in (**c**); the white rectangle shows the extent of (**a–c**).

2.3.2. Object Identification and Accuracy Assessment

The first step of object identification was to collect training objects. We assigned land cover classes to the objects fully containing the training samples based on their land cover types. In the second step, we constructed a feature space by making reference to literature reviews, expert knowledge, and visual examination. The feature space was composed of spectral, texture, and shape features, as described in Table 1. For example, the mean NDVI, NDWI, and LSWI values of all the pixels in an object were calculated, and were further used in the process of classification. In addition, DEM and ETOPO1 data were used to set the threshold for a specific region where the terrain feature should be considered. Generally, the thresholds for the coasts in different regions were different. For example, the value of two meters for DEM was used over the coast of the Dandou Sea. Next, the SVM classifier was trained with the collected training objects, the constructed feature space, and the algorithm parameters (radial basis function, RBF kernel) [38]. In this process, we visually compared the classification results from multiple groups of the RBF parameters, and found that the default values (C = 2 and γ = 0) of RBF in eCognition software are optimal for the SVM classifier. We then applied the trained SVM classifier to obtain an initial land cover classification. Subsequently, manual editing was performed to correct some misclassifications based on previous knowledge and field survey data, especially for patches near the boundaries of different vegetation types.

Final classification results was assessed using ground truth samples. A confusion matrix consisting of the overall accuracy, user accuracy, producer accuracy, and kappa coefficient was created to measure the consistency between our classification results and the validation samples. The generated results were used to construct a new dataset on the *S. alterniflora* invasion, which was called the Chinese Academy of Sciences *S. alterniflora* dataset (CAS *S. alterniflora*).

Table 1. Description of the feature space constructed for image classification.

No.	Feature	Attribute	Calculation Formula	Description
1	Mean value of each band	Spectral feature	$C_L = \frac{1}{n} \sum\limits_{i=1}^{n} C_{Li}$	C_{Li} represents the value of pixel i in band L, n is the number of pixels constructing an object, $i = 1, 2, \cdots n$, n_L is the number of bands and $L = 1, 2, \cdots n_L$, $\overline{C}_L$ is the mean value of each band
2	Brightness	Spectral feature	$Brightness = \frac{1}{n_L} \sum\limits_{i=1}^{n_L} \overline{C}_L$	
3	Standard deviation of each band	Spectral feature	$Stdv_L = \sqrt{\frac{1}{n-1} \cdot \sum\limits_{i=1}^{n} (C_{Li} - \overline{C}_L)^2}$	
4	GLCM homogeneity	Texture feature	$Homogeneity = \sum\limits_{i,j=0}^{N-1} \frac{P_{ij}}{1+(i-j)^2}$	P_{ij} denotes element i, j of the normalized symmetrical GLCM, and N is the number of gray levels in the image. Homogeneity is a feature related to the heterogeneity of pixels within an object. The values range from 0 to 1, and a higher value indicates a smoother texture feature.
5	Length–width ratio	Shape feature	Length-width ratio = Length /Width	The length-width ratio is useful for extracting linear features such as roads, dikes, and ditches.
6	Shape index	Shape feature	$SI = \frac{P}{4 \cdot \sqrt{A}}$	P is the object perimeter, and A is the object area.
7	NDVI	Spectral index	$NDVI = \frac{NIR - Red}{NIR + Red}$	NDVI utilizes the differential reflection of green vegetation in the red and near-infrared (NIR) portion to characterize vegetation condition.
8	NDWI	Spectral index	$NDWI = \frac{Green - NIR}{Green + NIR}$	The NDWI value of water is positive. In contrast, soil and vegetation on the ground have zero or negative NDWI values.
9	LSWI	Spectral index	$LSWI = \frac{NIR - SWIR1}{NIR + SWIR1}$	LSWI is sensitive to the total amount of liquid water in vegetation and the soil background.

3. Results

3.1. The Spatial Pattern of S. alterniflora in the Coastal Zone of Mainland China

The performed classification resulted in an overall accuracy of 96% and a kappa coefficient of 0.86, and producer and user accuracies greater than 0.85 (Table 2). These accurate classification results gave us confidence to describe the spatial pattern of *S. alterniflora* in the coastal zone of mainland China.

S. alterniflora was estimated to cover 545.80 km², and was found along the shoreline from the Nanpu coast in Tangshan, Hebei Province to Dafengjiang Estuary, Guangxi Province with a latitude from 39°13′N to 20°55′N. The spread of *S. alterniflora* is commonly by vegetative propagation after its artificial planting; *S. alterniflora* was thus found to present in clusters in most of the intertidal zones and estuaries of Jiangsu, Shanghai, Zhejiang, and Fujian provinces, and occupied a total area of 500.21 km² in these regions, accounting for nearly 92% of the total area of *S. alterniflora* in mainland China (Figure 5). However, *S. alterniflora* was scarce in the other five coastal provinces: Hebei, Tianjin, Shandong, Guangdong, and Guangxi, and was not observed in the northernmost coastal province, Liaoning.

Table 2. Confusion matrix of *S. alterniflora* classification result in the coastal zone of Mainland China.

Field Survey Points	Classification Result		
	S. alterniflora	Non-*S. alterniflora*	In Total
S. alterniflora	467	59	515
Non-*S. alterniflora*	76	2735	2811
Total	543	2794	3337
Producer accuracy	91%	97%	
User accuracy	86%	98%	
Overall accuracy	96%		
Kappa coefficient	0.86		

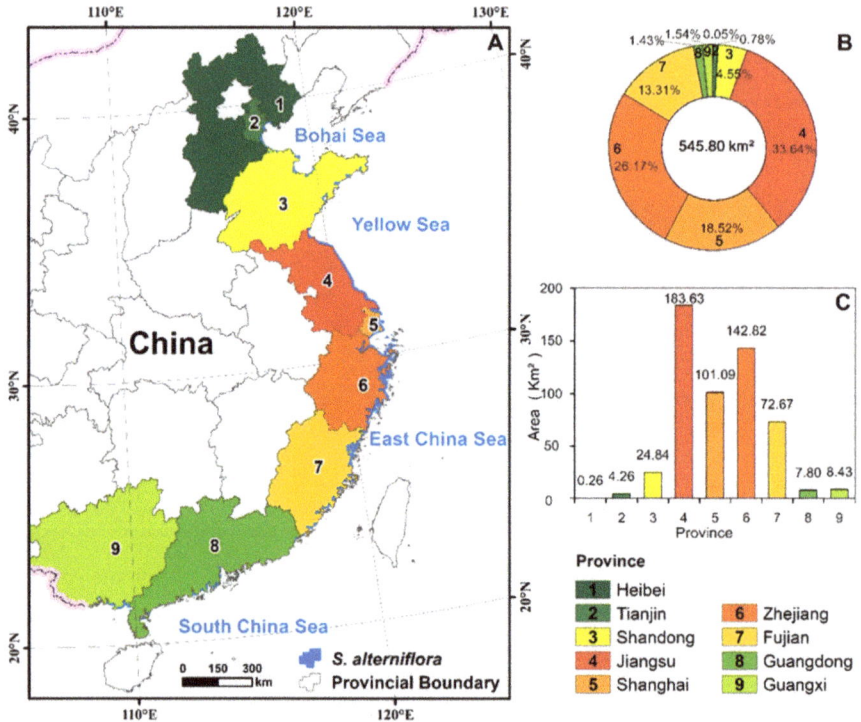

Figure 5. Spatial distribution (**A**), areal proportion (**B**), and total area (**C**) of *S. alterniflora* in the coastal provinces of mainland China in 2015.

3.2. Geospatially Varied Distributions of S. alterniflora in Coastal Provinces

Figure 6 illustrates the geospatially varied distributions of *S. alterniflora* in hotspot regions. *S. alterniflora* was found to be 0.26 km^2 in Hebei. Most of the *S. alterniflora* was distributed in patches along riverbanks, and grew parallel to the shoreline of Huanghua City. *S. alterniflora* in Tianjin was sporadically distributed from the Hangu coast in the north to Ziya River Estuary in the south, whereas the largest area of *S. alterniflora* was identified in Ziya River Estuary. *S. alterniflora* in Shandong was determined to be 24.84 km^2, and was mainly observed in the estuaries of this province such as the Yellow River Delta, Xiaoqing River Estuary, Dingzi Bay, Laizhou Bay, Jiaozhou Bay, and Rushan Bay in the areal order from large to small.

Jiangsu suffered the greatest invasion of *S. alterniflora* among all of the coastal provinces. The area of *S. alterniflora* in Jiangsu was estimated to be 183.63 km^2, accounting for 33.64% of the total invasion area in mainland China. *S. alterniflora* extended from the Xiuzhen River Estuary in the north to the

Qidong coast in the south, mainly in the intertidal zones of Dafeng and Rudong counties. Almost all of the major ports and estuaries were invaded by *S. alterniflora*. In Shanghai, the exotic plant was mainly identified in the northeast part of Chongming Island and Jiuduansha Shoals, and also found as narrow strips along the Nanhui coast. In Zhejiang Province, a considerable proportion of *S. alterniflora* was detected in the bay areas and major ports, with Sanmen Bay having the largest—32.21 km² —and Yueqing Bay having the second largest with 25.13 km². Ningbo had the largest area of *S. alterniflora* (74.61 km²) among the prefecture-level cities, accounting for over half of the total area in Zhejiang Province, followed by Taizhou and Wenzhou cities. Additionally, some patches of *S. alterniflora* were found in the coastal reclamation districts. The distribution of *S. alterniflora* in Fujian Province extended from Yacheng Bay in the north to Zhangjiang Estuary in the south, and covered most of the main estuaries and bay areas. Sandu Bay showed the largest areal extent of *S. alterniflora* (33.28 km²), representing 45.80% of the total invasion area of Fujian Province, followed by Luoyuan Bay (8.60 km²), Quanzhou Bay (7.53 km²), Minjiang Estuary (3.25 k km²), and Funing Bay (2.38 km²).

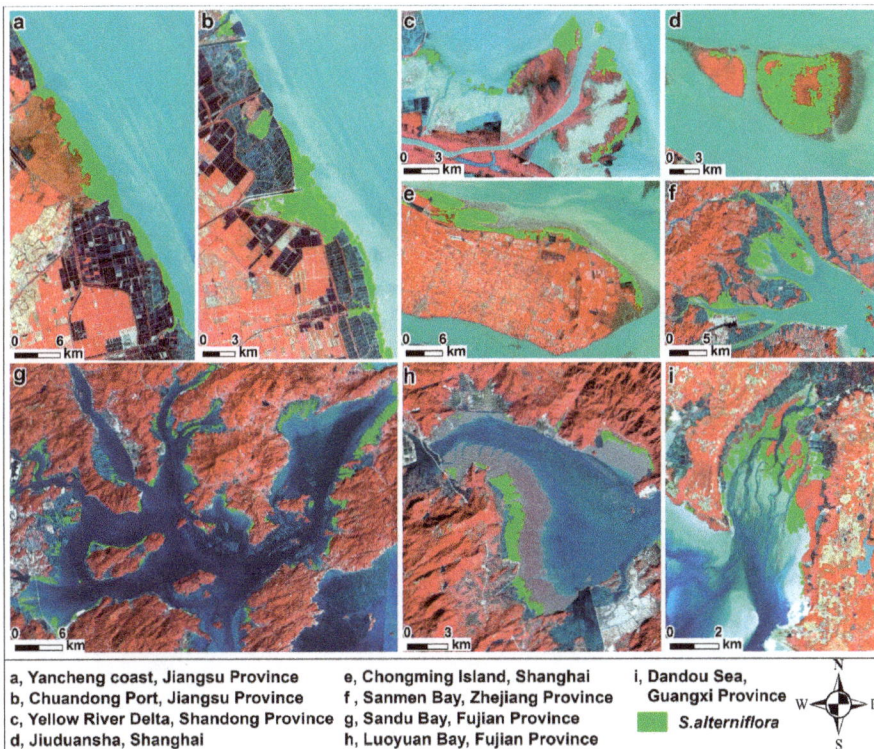

a, Yancheng coast, Jiangsu Province	e, Chongming Island, Shanghai
b, Chuandong Port, Jiangsu Province	f, Sanmen Bay, Zhejiang Province
c, Yellow River Delta, Shandong Province	g, Sandu Bay, Fujian Province
d, Jiuduansha, Shanghai	h, Luoyuan Bay, Fujian Province

i, Dandou Sea, Guangxi Province

S.alterniflora

Figure 6. *S. alterniflora* invasion over the hotspot regions in coastal provinces of Mainland China (Landsat-8 operational land imager (OLI) color combination: band 5 = red, band 4 = green, and band 3 = blue).

In Guangdong Province, most of the patches of *S. alterniflora* were observed along the shoreline and in the estuaries of Taishan and Zhuhai. The northernmost location covered by *S. alterniflora* was Yifengxi River Estuary, whereas the southernmost *S. alterniflora* patches were distributed along the Beishangang coast in Zhanjiang City. In Guangxi Province, the area of *S. alterniflora* was estimated to be 8.43 km², and the distribution of this exotic plant was concentrated in Yingluo Bay, followed by the Shatian coast, Dandou Sea, Tieshan Port, Yingpan Port, Lianzhou Bay, Nanliujiang Estuary, and Dafengjiang Estuary.

3.3. S. Alterniflora Invasion in Coastal NNRs

To specifically investigate the invasion of *S. alterniflora* to native ecosystems, the distribution of *S. alterniflora* within the coastal NNRs of mainland China was identified (Figure 7). Seven NNRs were markedly invaded by *S. alterniflora*: the Yellow River Delta NNR (YRDNNR), Yancheng NNR (YNNR), Chongming Dongtan NNR (CDNNR), Jiuduansha Wetland NNR (JWNNR), Zhangjiangkou Mangrove NNR (ZMNNR), Shankou Mangrove NNR (SMNNR), and Hepu Dugong NNR (HDNNR). A total area of 174.35 km² of *S. alterniflora* was mapped in these seven NNRs, accounting for 31.9% of the total area of *S. alterniflora* in mainland China. The area and proportional area of *S. alterniflora* were calculated for each NNR with respect to the different functional zones, varied significantly among the NNRs. Overall, the experimental zone had the largest area of *S. alterniflora* (71.39 km²), while the core zone displayed the highest coverage of *S. alterniflora* (5.25%).

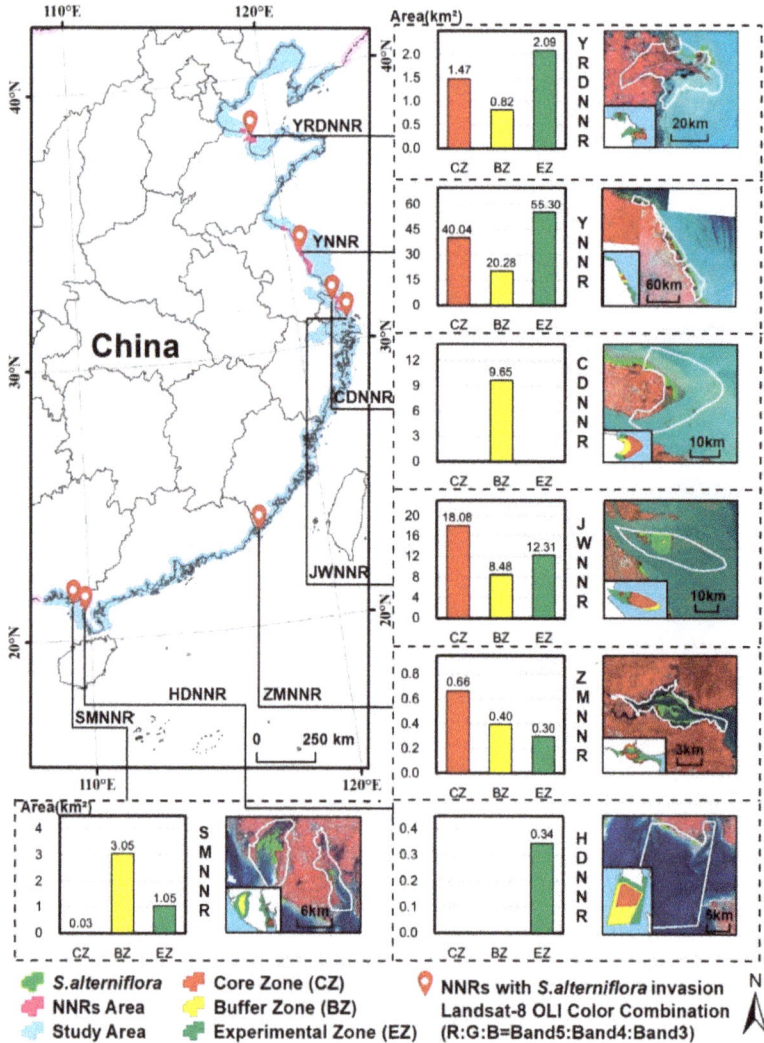

Figure 7. Spatial distribution of *S. alterniflora* and statistics of invasive areas at different functional zones within these coastal national nature reserves (NNRs) of Mainland China in 2015.

The YRDNNR was the northernmost NNR invaded by *S. alterniflora*. There was 4.38 km^2 of *S. alterniflora* dispersed along the seaward boundary of the intertidal zone in the southern part, and the experimental zone had the largest area of *S. alterniflora*. The YNNR, which was designed for protecting rare waterfowls, had the widest distribution of *S. alterniflora* with an area of 115.62 km^2, accounting for 66.31% of the total area of *S. alterniflora* in the NNRs. In this reserve, the species was distributed in strips almost parallel to the shoreline, and occupied the largest area in the experimental zone, followed by the core zone and buffer zone. In the CDNNR, 9.65 km^2 of *S. alterniflora* occurred as a strip in the intertidal mudflat of the buffer zone, whereas *S. alterniflora* were not observed in the core zone and experimental zone. Approximately 9.35% of the JWNNR was covered by *S. alterniflora*, which was mainly distributed in the core zone, followed by the experimental and buffer zones. In the ZMNNR, 1.36 km^2 of *S. alterniflora* was dispersed along the riverbanks as well as in the intertidal mudflats and shoals, and the invasive plant formed a strip in the southeast–northwest direction. *S. alterniflora* in the SMNNR was patchy in the mudflats of the Dandou Sea, occupying the largest area in the buffer zone, followed by the experimental zone. The area of *S. alterniflora* in the HDNNR was less than 1 km^2, and the invasive plant was patchy along the coast of the experimental zone.

4. Discussion

4.1. Landsat-Based Detection of S. Alterniflora Invasion

Remote sensing has been widely used in previous studies of *S. alterniflora* invasion to observe population development, detect spatiotemporal patterns, and characterize landscape dynamics [11,14,39,40]. However, an up-to-date dataset of *S. alterniflora* invasion at the national scale has been lacking in China. The CAS *S. alterniflora* dataset developed in this study achieved an updated and reliable mapping result of *S. alterniflora* invasion in mainland China from multiple aspects. This dataset documents the newest areas and current distribution of *S. alterniflora* (2015), the knowledge of which is crucial for dealing with the rapid and extensive expansion of *S. alterniflora* 40 years after its introduction. The OLI images from the newly launched Landsat 8 ensured improved mapping results superior to previous ones due to the greater number of spectral bands, superior spectral information, and greater availability of images compared with previous data sources with moderate spatial resolution [26]. In addition, Landsat series' satellites can provide long-term images to reconstruct the historical patterns of *S. alterniflora*, which can ensure data consistency. Furthermore, the OBIA method presents great advantages with respect to utilizing textural, geometric, and contextual features, avoiding salt-and-pepper noise, and accordingly improving classification accuracy and efficiency [32,33,41]. The FbSP optimizer developed for automatically determining optimal segmentation parameters can improve segmentation accuracy and reduce the operation time, and it is operator-independent [36,37]. The SVM classifier provides advantages for OBIA because the number of object samples tend to be fewer than that used by pixel-based approaches [42], and generally achieves higher classification accuracy than other traditional classification methods [11,43,44]. This combination is very effective for the classification of *S. alterniflora*. Additionally, a large number of training and validation samples from field surveys, which covered the whole coast of mainland China, greatly ensured the classification accuracy. Repetitive manual interpretation and comparison with previous reports at various regions and scales also contributed greatly to the reliability of this dataset.

Due to the variation in data sources, classification methods, and dataset dates among previous studies, there are many uncertainties in the assessments of the invasion mechanism and rates of *S. alterniflora*. Thus, it is necessary and important to develop multi-temporal datasets to continuously characterize the historical patterns and processes of *S. alterniflora* invasion. The approach used in this study is suitable to be generalized to build such a database. Remote sensing data of moderate spatial resolution are of limited utility for the detection of objects at fine or detailed scales. Definitely for *S. alterniflora*, Landsat images are of limited value for delimitating small and narrow patches due to their spectral uniformity, e.g., areas smaller than 1000 m^2, especially where *S. alterniflora* has not gained dominance [3,11]. High-resolution satellite data or the fusion of multiple data sources that cover the long coast of mainland China, combined with new classification methods such as machine learning, are thus needed to be assessed for a more accurate monitoring of *S. alterniflora* invasion.

4.2. Expansion Dynamics of S. Alterniflora

Monitoring the distribution of *S. alterniflora* has received extensive attention in China [3,14,23,45,46]. For mainland China, there were three studies investigated the distribution of *S. alterniflora* along the coast [10,12,14]. Specifically, two studies reported the estimated area and distribution of *S. alterniflora* around 2007 over the coast of mainland China (Table 3). Zuo et al. [12] generated the first mapping results of the distribution of *S. alterniflora* around 2007 using Landsat Thematic Mapper (TM) and CBERS images and categorization threshold methods. Subsequently, Lu and Zhang [10] investigated the spatial distribution of *S. alterniflora* around 2007 again based on CBERS images and a combination of supervised classification and visual interpretation. These two studies obtained similar results in the area of *S. alterniflora*. However, there was a pronounced difference between these studies in the distribution of *S. alterniflora*; the former study found *S. alterniflora* in Liaoning Province, whereas the latter did not (Table 3). Recently, Zhang et al. [14] examined the temporal change of *S. alterniflora* and identified *S. alterniflora* in Huludao, Liaoning Province, which was not validated by field investigation. Our mapping results confirmed that there was no *S. alterniflora* invasion in Liaoning as of 2015. During our field surveys, we found only some small patches of *Spartina anglica* (*S. anglica*) along the coasts of Jinzhou and Xingcheng in this province (Figure 8). The spectral and phenological similarity between *S. alterniflora* and *S. anglica* may led to the misclassification of *S. alterniflora* in Liaoning. In the future, it is necessary to investigate the possibility for the accurately differentiating *S. alterniflora* from *S. anglica* by the fusion of Landsat images and other data sources with finer resolution, such as Sentinel-2 or hyperspectral data [47,48].

The spread of *S. alterniflora* is commonly by vegetative propagation after its artificial planting, which makes the *S. alterniflora* present in clusters in most of the coasts. In this study, the total area of *S. alterniflora* was estimated to be 545.80 km^2, which indicates a mean expansion rate of 137 km^2 per decade from its introduction in China. Our finding and the previous estimates at the mainland China scale suggest that *S. alterniflora* expanded rapidly over a total area greater than 200 km^2 during the intervening decade. We also observed a northward expansion of *S. alterniflora* in mainland China. In our study, we identified *S. alterniflora* in northern Hebei, whereas the northern limit of the distribution reported by Lu and Zhang [10] was in Tianjin. Previous studies documented that *S. alterniflora* has strong adaptability in a variety of substrates [8,9]. Thus, although artificial planting has played a role, climate warming is probably the main driving force for the northward expansion of *S. alterniflora*, as the warming temperature meeting the ecological niche requirements for *S. alterniflora* growth. Considering the apparent consequences of *S. alterniflora* invasion in southern areas and the warming climate, there is a need to respond to this plant invasion in Hebei, and even in Liaoning Province, in spite of no *S. alterniflora* being identified at present.

Table 3. Characteristics of *S. alterniflora* distribution at the scale of mainland China from this study and previous studies.

Datasets	Estimated Area (km^2)	Spatial Extents	Data Source	Dataset Date
Zuo et al., 2012 [12]	344.51	>40°N–~21°27′N	Landsat TM & CBERS	2007
Lu and Zhang, 2013 [10]	341.78	39°05′N–21°27′N	CBERS	2007
Zhang et al., 2017 [14]	551.81	40°47′N–19°46′N	Landsat TM/ETM+	2014
This study	545.80	39°13′N–20°55′N	Landsat 8 OLI	2015

Figure 8. Photos for S. anglica observed along the coast in Jinzhou (**a1,a2**) and Xingcheng (**b1,b2**), Liaoning Province, and the *S. alterniflora* over the coast of Tangshan, Hebei Province (**c1,c2**) from the field investigation in 2016.

4.3. Potential Effects of S. alterniflora Invasion

S. alterniflora was originally introduced in China to protect dikes and promote silting for land reclamation [9,11,12]. In some areas along China's coast, *S. alterniflora* played vital roles in achieving these goals. The deep root systems and high salt and wave tolerance of *S. alterniflora* have greatly reduced the influences of wind waves and typhoons along the coast. For example, the distribution of *S. alterniflora* over the coast of Winzhou of Zhejiang had significantly protected the coastal environment against typhoon in 1990 and 1994 [9]. Moreover, *S. alterniflora* has apparently contributed to land reclamation in some areas, especially on the coast of Jiangsu Province, where the shoreline has obviously extended seaward [49]. The high biomass and coverage of *S. alterniflora* not only significantly traps the sediment from seawater, but also has great potential for carbon sequestration and the production of animal fodder and biofuels [9,10,50]. In addition, *S. alterniflora* provides important shelter and food for many terrestrial animal, waterfowl, and fish communities [51,52]. Given the high productivity, extensive distribution, and rapid expansion of *S. alterniflora*, it is worth deeply studying the scientific utilization *S. alterniflora*.

Even so, the negative consequences of *S. alterniflora* expansion were being increasingly recognized. High-density areas of *S. alterniflora* can cause microtopographical changes in ports and block waterways. Furthermore, they can threaten coastal water quality by affecting the exchange capacity of seawater and impede coastal economic development [9]. Thus, effective control of the *S. alterniflora* invasion in such regions is necessary. Although the extensive areas of *S. alterniflora* have high carbon sequestration potential, they also have high levels of methane emission [14,53,54]. Therefore, their capacity to mitigate global warming requires further assessments. Owing to its high adaptability, *S. alterniflora* has encroached upon large mudflat areas, which has reduced the foraging habitat for waterfowls, such as the *Larus saundersi* [5,55]. Moreover, the invasive *S. alterniflora* has replaced numerous native plants, including *Phragmites australis*, *Suaeda glauca*, mangroves, and *Scirpus planiculmis*, which has affected ecosystem structures and processes [11,23], habitat suitability for endangered waterfowl [56,57], and regional tourism [9]. As found in our field investigation, the encroachment of *S. alterniflora* to *Suaeda glauca* noticeably affected the original beautiful landscape "red beach". Meanwhile, extensive *S. alterniflora* were identified in the NNRs (Figure 7), which were designed for protecting native species. Therefore, the potential effects of *S. alterniflora* invasion should be objectively evaluated at local scales to allow scientific and region-specific decisions to promote sustainable coastal ecosystems and economic development.

5. Conclusions

In this study, we have mapped the spatial distribution of *S. alterniflora* invasion in 2015 by applying OBIA and SVM approaches to multiple scenes of Landsat 8 OLI images over the coast zone of mainland China. The classification method and data source yielded reliable spatial information for *S. alterniflora* in 2015 with high accuracy, which was validated by a large number of ground truth samples. This dataset and related analyses are expected to guide scientific management regarding *S. alterniflora* invasion to promote sustainable coastal ecosystems. The up-to-date observation revealed that the total area of *S. alterniflora* was about 545.80 km^2; this exotic species was identified from the Nanpu coast of Hebei in the northernmost region to Dafengjiang Estuary of Guangxi Province in the southernmost area. Nearly 92% of the total area of *S. alterniflora* was distributed within four provinces, including Jiangsu, Shanghai, Zhejiang, and Fujian (500.21 km^2), which need particular attention. In addition, seven of 32 NNRs that were established to protect native animal or plant species over the coast of mainland China have been markedly invaded by *S. alterniflora*, with the total area accounting for about one-third of the total invasion area in mainland China. Given the rapid expansion of *S. alterniflora* since its introduction and the serious ecological effects, effective response decisions are urgently needed.

Author Contributions: M.L. conceived and designed the research, processed the data, and wrote the manuscript draft. D.M. helped design the research and revise the manuscript. Z.W., L.L. and Y.Z. helped conceive the research and review the manuscript. W.M. conducted the fieldwork and analyzed the data. M.J. and C.R. contributed materials.

Funding: This study is jointly supported by the National Key R&D Program of China (2016YFC0500201), the Science &Technology Basic Resources Investigation Program of China (2017FY100706), the Strategic Priority Research Program of the Chinese Academy of Sciences (XDA19040503), and the National Natural Science Foundation of China (41771383).

Acknowledgments: We would like to thank the USGS for providing satellite images and thank the three anonymous reviewers for the constructive comments on our manuscript.

Conflicts of Interest: The authors declare no conflict of interest.

References

1. Alvarez-Taboada, F.; Paredes, C.; Julián-Pelaz, J. Mapping of the invasive species Hakea sericea using unmanned aerial vehicle (UAV) and WorldView-2 imagery and an object-oriented approach. *Remote Sens.* **2017**, *9*, 913. [CrossRef]
2. MacDougall, A.S.; Boucher, J.; Turkington, R.; Bradfield, G.E. Patterns of plant invasion along an environmental stress gradient. *J. Veg. Sci.* **2009**, *17*, 47–56. [CrossRef]
3. Liu, C.; Jiang, H.; Zhang, S.; Li, C.; Pan, X.; Lu, J.; Hou, Y. Expansion and Management Implications of Invasive Alien Spartina alterniflora in Yancheng Salt Marshes, China. *Open J. Ecol.* **2016**, *6*, 113–128. [CrossRef]
4. Mack, R.N.; Simberloff, D.; Lonsdale, W.M.; Evans, H.; Clout, M.; Bazzaz, F.A. Biotic invasions: Causes, epidemiology, global consequences, and control. *Ecol. Appl.* **2000**, *10*, 689–710. [CrossRef]
5. Gan, X.; Cai, Y.; Choi, C.Y.; Ma, Z.; Chen, J.; Li, B. Potential impacts of invasive Spartina alterniflora on spring bird communities at Chongming Dongtan, a Chinese wetland of international importance. *Estuar. Coast. Shelf Sci.* **2009**, *83*, 211–218. [CrossRef]
6. Levin, L.A.; Neira, C.; Grosholz, E.D. Invasive cordgrass modifies wetland trophic function. *Ecology* **2006**, *87*, 419–432. [CrossRef] [PubMed]
7. Zhang, Y.; Huang, G.; Wang, W.; Chen, L.; Lin, G. Interactions between mangroves and exotic Spartina in an anthropogenically disturbed estuary in southern China. *Ecology* **2012**, *93*, 588–597. [CrossRef] [PubMed]
8. An, S.; Gu, B.; Zhou, C.; Wang, Z.; Deng, Z.; Zhi, Y.; Li, H.; Chen, L.; Yu, D.; Liu, Y. Spartina invasion in China: Implications for invasive species management and future research. *Weed Res.* **2007**, *47*, 183–191. [CrossRef]
9. Chung, C.H. Forty years of ecological engineering with Spartina plantations in China. *Ecol. Eng.* **2006**, *27*, 49–57. [CrossRef]
10. Lu, J.; Zhang, Y. Spatial distribution of an invasive plant Spartina alterniflora and its potential as biofuels in China. *Ecol. Eng.* **2013**, *52*, 175–181. [CrossRef]
11. Wang, A.; Chen, J.; Jing, C.; Ye, G.; Wu, J.; Huang, Z.; Zhou, C. Monitoring the invasion of Spartina alterniflora from 1993 to 2014 with Landsat TM and SPOT 6 Satellite Data in Yueqing Bay, China. *PLoS ONE* **2015**, *10*, e0135538. [CrossRef] [PubMed]
12. Zuo, P.; Zhao, S.; Liu, C.; Wang, C.; Liang, Y. Distribution of Spartina spp. along China's coast. *Ecol. Eng.* **2012**, *40*, 160–166. [CrossRef]
13. Jia, M.; Wang, Z.; Zhang, Y.; Mao, D.; Wang, C. Monitoring loss and recovery of mangrove forests during 42 years: The achievements of mangrove conservation in China. *Int. J. Appl. Obs.* **2018**, *73*, 535–545. [CrossRef]
14. Zhang, D.; Hu, Y.; Liu, M.; Chang, Y.; Yan, X.; Bu, R.; Zhao, D.; Li, Z. Introduction and spread of an exotic plant, Spartina alterniflora, along coastal marshes of China. *Wetlands* **2017**, *37*, 1181–1193. [CrossRef]
15. Chen, Y.; Chen, G.; Ye, Y. Coastal vegetation invasion increases greenhouse gas emission from wetland soils but also increases soil carbon accumulation. *Sci. Total Environ.* **2015**, *526*, 19–28. [CrossRef] [PubMed]
16. Bradley, B.A. Remote detection of invasive plants: A review of spectral, textural and phenological approaches. *Biol. Invasions* **2014**, *16*, 1411–1425. [CrossRef]
17. Lawrence, R.L.; Wood, S.D.; Sheley, R.L. Mapping invasive plants using hyperspectral imagery and Breiman Cutler classifications (Random Forest). *Remote Sens. Environ.* **2006**, *100*, 356–362. [CrossRef]
18. Niphadkar, M.; Nagendra, H. Remote sensing of invasive plants: Incorporating functional traits into the picture. *Int. J. Remote Sens.* **2016**, *37*, 3074–3085. [CrossRef]
19. Underwood, E.; Ustin, S.; DiPietro, D. Mapping nonnative plants using hyperspectral imagery. *Remote Sens. Environ.* **2003**, *86*, 150–161. [CrossRef]
20. Huang, C.; Asner, G.P. Applications of remote sensing to alien invasive plant studies. *Sensors* **2009**, *9*, 4869–4889. [CrossRef]
21. Müllerová, J.; Pergl, J.; Pyšek, P. Remote sensing as a tool for monitoring plant invasions: Testing the effects of data resolution and image classification approach on the detection of a model plant species Heracleum mantegazzianum (giant hogweed). *Int. J. Appl. Obs.* **2013**, *25*, 55–65. [CrossRef]
22. Civille, J.C.; Sayce, K.; Smith, S.D.; Strong, D.R. Reconstructing a century of Spartina alterniflora invasion with historical records and contemporary remote sensing. *Ecoscience* **2005**, *12*, 330–338. [CrossRef]

23. Wan, H.; Wang, Q.; Jiang, D.; Fu, J.; Yang, Y.; Liu, X. Monitoring the invasion of Spartina alterniflora using very high resolution unmanned aerial vehicle imagery in Beihai, Guangxi (China). *Sci. World J.* **2014**, *2014*, 638296. [CrossRef] [PubMed]

24. Liu, M.; Li, H.; Li, L.; Man, W.; Jia, M.; Wang, Z.; Lu, C. Monitoring the invasion of spartina alterniflora using multi-source high-resolution imagery in the Zhangjiang Estuary, China. *Remote Sens.* **2017**, *9*, 539. [CrossRef]

25. O'Donnell, J.; Schalles, J. Examination of abiotic drivers and their influence on Spartina alterniflora biomass over a twenty-eight year period using Landsat 5 TM Satellite Imagery of the Central Georgia Coast. *Remote Sens.* **2016**, *8*, 477. [CrossRef]

26. Roy, D.P.; Wulder, M.A.; Loveland, T.R.; Woodcock, C.E.; Allen, R.G.; Anderson, M.C.; Helder, D.; Irons, J.R.; Johnson, D.M.; Kennedy, R.; et al. Landsat-8: Science and product vision for terrestrial global change research. *Remote Sens. Environ.* **2014**, *145*, 154–172. [CrossRef]

27. Ma, Z.; Melville, D.S.; Liu, J.; Chen, Y.; Yang, H.; Ren, W.; Zhang, Z.; Piersma, T.; Li, B. Rethinking China's new great wall. *Science* **2014**, *346*, 912–914. [CrossRef] [PubMed]

28. Ouyang, Z.; Zhang, M.; Xie, X.; Shen, Q.; Guo, H.; Zhao, B. A comparison of pixel-based and object-oriented approaches to VHR imagery for mapping saltmarsh plants. *Ecol. Inform.* **2011**, *6*, 136–146. [CrossRef]

29. Yu, Q.; Gong, P.; Clinton, N.; Biging, G.; Kelly, M.; Schirokauer, D. Object-based detailed vegetation classification with airborne high spatial resolution remote sensing imagery. *Photogramm. Eng. Rem. Sci.* **2006**, *72*, 799–811. [CrossRef]

30. Ouyang, Z.; Gao, Y.; Xie, X.; Guo, H.; Zhang, T.; Zhao, B. Spectral discrimination of the invasive plant spartina alterniflora at multiple phenological stages in a saltmarsh wetland. *PLoS ONE* **2013**, *8*, e67315. [CrossRef] [PubMed]

31. Gao, Z.; Zhang, L. Multi-seasonal spectral characteristics analysis of coastal salt marsh vegetation in Shanghai, China. *Estuar. Coast. Shelf Sci.* **2006**, *69*, 217–224. [CrossRef]

32. Dronova, I. Object-based image analysis in wetland research: A review. *Remote Sens.* **2015**, *7*, 6380–6413. [CrossRef]

33. Jawak, S.D.; Devliyal, P.; Luis, A.J. A comprehensive review on pixel oriented and object oriented methods for information extraction from remotely sensed satellite images with a special emphasis on cryospheric applications. *Adv. Remote Sens.* **2015**, *4*, 177–195. [CrossRef]

34. Mountrakis, G.; Im, J.; Ogole, C. Support vector machines in remote sensing: A review. *ISPRS J. Photogramm.* **2011**, *66*, 247–259. [CrossRef]

35. Pal, M.; Mather, P.M. Support vector machines for classification in remote sensing. *Int. J. Remote Sens.* **2005**, *26*, 1007–1011. [CrossRef]

36. Tong, H.; Maxwell, T.; Zhang, Y.; Dey, V. A supervised and fuzzy-based approach to determine optimal multi-resolution image segmentation parameters. *Photogramm. Eng. Rem. Sci.* **2012**, *78*, 1029–1044. [CrossRef]

37. Zhang, Y.; Maxwell, T.; Tong, H.; Dey, V. Development of supervised software tool for automated determination of optimal segmentation parameters for eCognition. In Proceedings of the ISPRS TC VII symposium-100 Years ISPRS, Vienna, Austria, 5–7 July 2010.

38. Huang, C.; Davis, L.S.; Townshend, J.R.G. An assessment of support vector machines for land cover classification. *Int. J. Remote Sens.* **2002**, *23*, 725–749. [CrossRef]

39. Huang, H.; Zhang, L. A study of the population dynamics of Spartina alterniflora at Jiuduansha shoals, Shanghai, China. *Ecol. Eng.* **2007**, *29*, 164–172. [CrossRef]

40. Long, X.; Liu, L.; Shao, T.; Shao, H.; Liu, Z. Developing and sustainably utilize the coastal mudflat areas in China. *Sci. Total Environ.* **2016**, *569*, 1077–1086. [CrossRef]

41. Whiteside, T.G.; Boggs, G.S.; Maier, S.W. Comparing object-based and pixel-based classifications for mapping savannas. *Int. J. Appl. Obs.* **2011**, *13*, 884–893. [CrossRef]

42. Tzotsos, A.; Argialas, D.A. Support vector machine approach for object based image analysis. In *Object-Based Image Analysis*; Springer: Berlin/Heidelberg, Germany, 2008; pp. 663–677.

43. Heumann, B.W. An object-based classification of mangroves using a hybrid decision tree-support vector machine approach. *Remote Sens.* **2011**, *3*, 2440–2460. [CrossRef]

44. Mantero, P.; Moser, G.; Serpico, S.B. Partially supervised classification of remote sensing images through SVM-based probability density estimation. *IEEE T. Geosci. Remote Sens.* **2005**, *43*, 559–570. [CrossRef]

45. Liu, X.; Liu, H.; Gong, H.; Lin, Z.; Lv, S. Appling the one-class classification method of maxent to detect an invasive plant Spartina alterniflora with time-series analysis. *Remote Sens.* **2017**, *9*, 1120. [CrossRef]

46. Zhang, W.; Zeng, C.; Tong, C.; Zhang, Z.; Huang, J. Analysis of the expanding process of the Spartina alterniflora salt marsh in Shanyutan wetland, Minjiang River Estuary by remote sensing. *Procedia Environ. Sci.* **2011**, *10*, 2472–2477. [CrossRef]

47. Griffiths, P.; Nendel, C.; Hostert, P. Intra-annual reflectance composites from Sentinel-2 and Landsat for national-scale crop and land cover mapping. *Remote Sens. Environ.* **2019**, *220*, 135–151. [CrossRef]

48. Villa, P.; Pinardi, M.; Bolpagni, R.; Gillier, J.; Zinke, P.; Nedelcut, F.; Bresciani, M. Assessing macrophyte seasonal dynamics using dense time series of medium resolution satellite data. *Remote Sens. Environ.* **2018**, *216*, 230–244. [CrossRef]

49. Hou, X.; Wu, T.; Hou, W.; Chen, Q.; Wang, Y.; Yu, L. Characteristics of coastline changes in mainland China since the early 1940s. Sci. *China Earth Sci.* **2016**, *59*, 1791–1802. [CrossRef]

50. Liu, J.; Han, R.; Su, H.; Wu, Y.; Zhang, L.; Richarson, C.J.; Wang, G. Effects of exotic Spartina alterniflora on vertical soil organic carbon distribution and storage amount in coastal salt marshes in Jiangsu, China. *Ecol. Eng.* **2017**, *106*, 132–139. [CrossRef]

51. Cai, B.; He, Q.; An, Y. Spartina alterniflora invasions and effects on crab com-munities in a western Pacific estuary. *Ecol. Eng.* **2011**, *37*, 1920–1924. [CrossRef]

52. Feng, J.; Huang, Q.; Qi, F.; Guo, J.; Lin, G. Utilization of exotic Spartina alterniflora by fish community in the mangrove ecosystem of Zhangjiang Estuary: Evidence from stable isotope analyses. *Biol. Invasions* **2015**, *17*, 2113–2121. [CrossRef]

53. Gao, G.; Li, P.; Shen, Z.; Qin, Y.; Zhang, X.; Ghoto, K.; Zhu, X.; Zheng, H. Exotic Spartina alterniflora invasion increases CH_4 while reduces CO_2 emissions from mangrove wetland soils in southeastern China. *Sci. Rep.* **2018**, *8*, 9243. [CrossRef] [PubMed]

54. Xiang, J.; Liu, D.; Ding, W.; Yuan, J.; Lin, Y. Invasion chronosequence of Spartina alterniflora on methane emission and organic carbon sequestration in a coastal salt marsh. *Atmos. Environ.* **2015**, *112*, 72–80. [CrossRef]

55. Ma, Z.; Gan, X.; Cai, Y.; Chen, J.; Li, B. Effects of exotic Spartina alterniflora on the habitat patch associations of breeding saltmarsh birds at Chongming Dongtan in the Yangtze River estury, China. *Biol. Invasions* **2011**, *13*, 1673–1686. [CrossRef]

56. Liu, C.; Jiang, H.; Hou, Y.; Zhang, S.; Su, L.; Li, X.; Pan, X.; Wen, Z. Habitat changes for breeding waterbirds in Yancheng national nature reserve, China: A remote sensing study. *Wetlands* **2010**, *30*, 879–888. [CrossRef]

57. Tian, B.; Zhou, Y.; Zhang, L.; Yuan, L. Analyzing the habitat suitability for migratory birds at the Chongming Dongtan nature reserve in Shanghai, China. *Estuar. Coast. Shelf Sci.* **2008**, *80*, 296–302. [CrossRef]

remote sensing

MDPI

Article

Mapping Substrate Types and Compositions in Shallow Streams

Milad Niroumand-Jadidi [1,2,3,]*, Nima Pahlevan [4,5] and Alfonso Vitti [1]

1 Department of Civil, Environmental, and Mechanical Engineering, University of Trento, Via Mesiano 77, 38123 Trento, Italy; alfonso.vitti@unitn.it
2 Department of Biology, Chemistry, and Pharmacy, Freie Universität Berlin, Altensteinstraße 6, 14195 Berlin, Germany
3 Leibniz-Institute of Freshwater Ecology and Inland Fisheries, Müggelseedamm 310, 12587 Berlin, Germany
4 NASA Goddard Space Flight Center, 8800 Greenbelt Road, Greenbelt, MD 20771, USA; nima.pahlevan@nasa.gov
5 Science Systems and Applications, Inc., 10210 Greenbelt Road, Suite 600, Lanham, MD 20706, USA
* Correspondence: m.niroumand@unitn.it; Tel.: +39-0461-314392

Received: 5 January 2019; Accepted: 24 January 2019; Published: 29 January 2019

Abstract: Remote sensing of riverbed compositions could enable advances in hydro-morphological and habitat modeling. Substrate mapping in fluvial systems has not received as much attention as in nearshore, optically shallow inland, and coastal waters. As finer spatial-resolution image data become more available, a need emerges to expand research on the remote sensing of riverbed composition. For instance, research to date has primarily been based on spectral reflectance data from above the water surface without accounting for attenuation by the water-column. This study analyzes the impacts of water-column correction for substrate mapping in shallow fluvial systems (depth < 1 m). To do so, we performed three different experiments: (a) analyzing spectroscopic measurements in a hydraulic laboratory setting, (b) simulating water-leaving radiances under various optical scenarios, and (c) evaluating the potential to map bottom composition from a WorldView-3 (WV3) image of a river in Northern Italy. Following the retrieval of depth and diffuse attenuation coefficient (K_d), bottom reflectances were estimated using a water-column correction method. The results indicated significant enhancements in streambed maps based on bottom reflectances relative to maps produced from above-water spectra. Accounting for deep-water reflectance, embedded in the water-column correction, was demonstrated to have the greatest impact on the retrieval of bottom reflectance in NIR bands, when the water column is relatively thick (>0.5 m) and/or when the water is turbid. We also found that the WV3's red-edge band (i.e., 724 nm) considerably improved the characterization of submerged aquatic vegetation (SAV) densities from either above-water or retrieved bottom spectra. This study further demonstrated the feasibility of mapping SAV density classes from a WV3 image of the Sarca River in Italy by retrieving the bottom reflectances.

Keywords: substrate; aquatic vegetation; bottom reflectance; water-column correction; river; spectroscopy; radiative transfer; WorldView-3

---

1. Introduction

Consistent, accurate, and timely information on riverbed conditions is critical for management of fluvial systems [1–4]. Bottom type/composition, along with the topography of the riverbed, affects flow and sediment transport and provides physical habitat [3,5]. For instance, submerged aquatic vegetation (SAV) plays a critical role in structuring ecological, morphological, and hydraulic conditions of riverine environments. SAV provides habitat for a wide range of aquatic fauna such as fish, waterfowl, shellfish, and invertebrates [6] and can be considered as an indicator of water quality and general stream

Remote Sens. **2019**, *11*, 262

health [3,7]. Moreover, accounting for the presence of SAV is of particular importance in hydraulic and morphodynamic modeling [8].

Conventional methods of collecting information on riverbed composition are costly, time-consuming, and spatially and temporally sparse [9–11]. Remote sensing approaches can provide an efficient means of characterizing fluvial systems across large spatial and temporal extents [2,12,13]. From a remote sensing point of view, a riverbed can be characterized based on its geometrical and spectral features. Applications of through-water photogrammetry techniques using aerial and close-range imagery have long been used to analyze riverbed geometry and topography [1]. More recently, bathymetric light detection and ranging (LiDAR) technology [14,15] and structure-from-motion (SfM) photogrammetry using unmanned aerial vehicles (UAVs) [2] have been incorporated into analyses of bed topography. However, spectral-based analysis has been mostly limited to bathymetry, so characterization of bottom types and compositions has not been fully explored in riverine environments [3,16].

In spite of a sound background established by coastal studies, remote sensing of bottom properties in the context of riverine systems requires significant additional research [3,10]. Supervised classification was used to map nuisance green algae using RGB images acquired by a UAV [17]. Object-based analyses were also used to discriminate submerged macrophyte species from very high-resolution terrestrial and UAV images [10]. Anker et al. [18] claimed that spatial resolution is more important than spectral resolution for mapping macrophyte cover in a small stream by comparing aerial digital photography and hyperspectral imagery (4 cm vs. 1 m spatial resolution, respectively). However, most previous research has been based upon above-water reflectance data that do not account for water-column attenuation [3]. The above-water reflectances/radiances are influenced by the attenuation of light through the water column that can be a limiting factor for characterization and classification of substrate types in optical imagery [10]. More recently, Legleiter et al. [3] examined the possibility of retrieving bottom reflectances by accounting for depth and attenuation effects. They measured the diffuse attenuation coefficient (K_d) directly in the field and then retrieved bottom reflectances to classify sediment facies and algal density in the Snake River (Wyoming, USA) from field spectra and airborne hyperspectral imagery. Their preliminary results indicated no improvements in riverbed classification accuracy using bottom reflectances rather than above-water reflectances. However, there is still a need to further investigate the effects of water-column attenuation in various optical conditions such as variable inherent optical properties (IOPs), bottom types, and water depths [3]. In general, the relatively coarse spatial resolution (i.e., 30 m) of publicly available satellite imagery has been a key barrier in studying fluvial systems [10,19]. With the increasing availability of high resolution satellite imagery, applications of satellites have recently been expanded to riverine environments as well [3,20–23]. With the private sector involved in Earth-imaging, such as Planet Labs and DigitalGlobe, high-resolution image data are more likely to be used more frequently for management and decision-making [24].

The primary goal of this manuscript is to perform a comprehensive analysis on the impacts of water-column correction for the remote sensing of bottom types in fluvial systems. In this context, three different experiments were conducted: analyzing spectroscopic measurements in a hydraulic laboratory setting, performing radiative transfer simulations, and evaluating WV3 imagery for mapping riverbed composition. Our objectives were to (1) examine an approach for estimation of K_d in shallow rivers using above-water reflectances over a range of in-situ/known depths, which enables water-column correction for the bottom reflectance retrieval, (2) assess bottom-type mapping and SAV retrievals before and after accounting for water-column attenuation, (3) examine the utility of WV3's eight visible-near-infrared (VNIR) bands compared to four-band GeoEye data for mapping substrate properties, and (4) characterize the areal density of SAVs using a WV3 image of the Sarca River in Northern Italy.

The following section introduces the study area and the datasets associated with the three experiments. The methodology of our study is introduced in Section 3. We then elaborate on the experiments and the corresponding results in Section 4. Section 5 includes an overall discussion of the results and the implications for substrate mapping in fluvial systems. The manuscript concludes in Section 6 with a summary of our investigations and a number of recommendations for future studies.

2. Study Area and Datasets

To perform an application-relevant analysis, the three experiments were designed based on the hydro-geomorphological and optical properties of the Sarca River. The Sarca River is a very shallow river in the Italian Alps supplied with meltwater from the Adamello glaciers and flowing into Lake Garda (Figure 1). The riverbed in the study area is composed of gravels (primarily dolomite) with patches of SAV. The mean channel width is about 30 m and the water depth <1 m with an average of about 0.5 m in the study region. The ranges of water column constituents are available from long-term measurements in the study area [25].

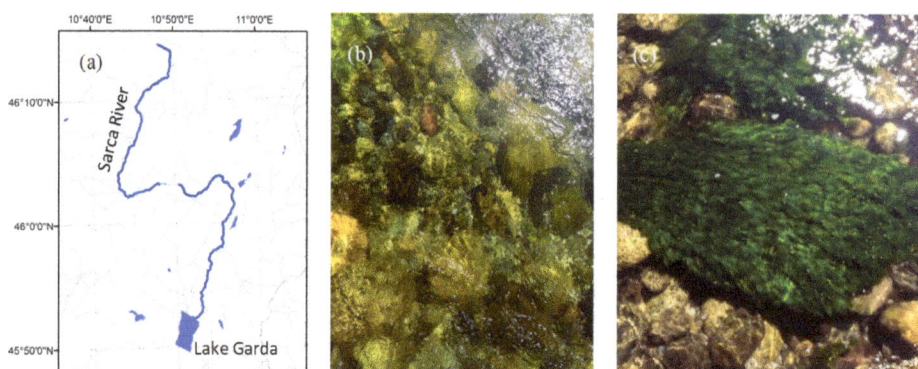

Figure 1. (a) The Sarca River located in Northeast Italy and main bottom types including (b) gravels composed of dolomite and (c) patches of submerged aquatic vegetation (SAV).

To perform a comprehensive assessment of the bottom reflectance retrieval methodology, this study applied three different radiometric datasets: one measured in a laboratory, one simulated using Hydrolight radiative transfer modeling [26], and one collected by the WV3 satellite sensor. The laboratory experiments allowed for controlled measurements of surface reflectance for flowing water with different SAV densities. The simulated spectra enabled an assessment of streambed mapping in a range of bottom types, water depths, and water column constituents representative of a wide range of optical conditions. The multispectral WV3 image of the Sarca River was also used to classify SAV densities to assess the feasibility and effectiveness of water-column correction from space. The measured and simulated reflectances were convolved with spectral responses of WV3 and GeoEye sensors (Table 1).

Table 1. Multispectral band designations for GeoEye and WV3 sensors [27].

GeoEye			WV3		
Band	Center Wavelength (nm)	Bandwidth (nm)	Band	Center Wavelength (nm)	Bandwidth (nm)
Blue (B)	484	76	Coastal-Blue (CB)	426	60
Green (G)	547	81	Blue (B)	481	72
Red (R)	676	42	Green (G)	547	79
NIR	851	156	Yellow (Y)	605	49
			Red (R)	661	70
			Red Edge (RE)	724	51
			NIR1	832	134
			NIR2	948	182

Table 2 provides a summary of datasets while more details regarding the experiments are provided in Section 4. The constituents are described in terms of concentrations of total suspended sediment (TSS), chlorophyll-a (Chl-a), and the absorption of colored dissolved organic mater at 440 nm (a_{CDOM} (440)).

Table 2. Datasets used in this study and their specifications.

Dataset	Spectral Characteristics	Bottom Types	Water Depths	Constituents
Laboratory	Spectroradiometric data with 1 nm resolution convolved to WV3 and GeoEye bands	Non-vegetated gravel, SAV with different densities	0 to 0.4 m with 1 cm intervals	Clear water with low TSS (~2 g/m³)
Synthetic	Hydrolight simulations with 10 nm resolution convolved to WV3 and GeoEye bands	Sediment, Macrophyte and Dolomite	0 to 1 m with 2 cm intervals	TSS = 2–6 g/m³ Chl-a = 1–5 mg/m³ a_{CDOM} (440) = 0.07–0.22 m⁻¹
Satellite	8-band WV3 image	SAV with different densities	0 to 0.8 m	TSS ~ 3 g/m³ Chl-a ~ 2 mg/m³ a_{CDOM} (440) ~ 0.09 m⁻¹

3. Methods

3.1. Water-Column Correction

The early work of Lyzenga [28,29] provided a physical basis for water-column correction and estimation of depth-invariant indices to map bottom properties in coastal settings. A review of bottom mapping techniques developed for remote sensing of coral reefs, algae, and seagrass is provided in [30]. Bottom mapping has been poorly studied in the context of fluvial systems and has mostly been based on above-water reflectances, which neglect the attenuation effects of the water column. The first attempt to apply existing water-column correction techniques in a riverine environment was the work by Legleiter et al. [3] based on limited, field-based spectral measurements. Their results demonstrated that sediment facies and algal densities can be characterized via their spectral information and suggested that retrieving bottom reflectances was not necessary. However, they indicated that the results were based on subjective interpretations of substrate images and suggested that more systematic studies, including radiative transfer modeling, would be needed to explore the potential for bottom reflectance retrieval. This research attempts to employ similar physics-based approaches to map bottom types using spectral reflectance data. In this study, the bottom reflectance retrieval is assisted by estimating K_d from image data using known water depths to eliminate the need for field-based spectral measurements carried out by the previous work [3].

The remote sensing reflectance (R_{rs}), defined as the ratio of the water-leaving radiance to the total downwelling irradiance just above the water surface, is an apparent optical property critical for analysis of optical imagery over water bodies [31–33]. Radiometric and atmospheric corrections are required to derive R_{rs} from top of atmosphere (TOA) radiances [32]. Note that reflectances/radiances and K_d are all wavelength (λ)-dependent; however, we drop λ for brevity in the text while retaining it in equations. Remote sensing reflectance just beneath the water surface (r_{rs}) can then be estimated to account for transmission and refraction at the air–water interface [3,34]:

$$r_{rs}(\lambda) = \frac{R_{rs}(\lambda)}{0.52 + 1.7R_{rs}(\lambda)}. \tag{1}$$

Thereafter, the remote sensing reflectance of bottom (r_{rs}^{B}) can be estimated according to the following equation [3,35,36]:

$$r_{rs}^{B}(\lambda) = \frac{r_{rs}(\lambda) - R_{rs}^{\infty}(\lambda)\left[1 - e^{-2K_d(\lambda)d}\right]}{e^{-2K_d(\lambda)d}} \tag{2}$$

where R_{rs}^{∞} denotes the remote sensing reflectance of optically deep water (i.e., negligible bottom-reflected radiance). The parameter K_d is the spectral diffuse attenuation coefficient that characterizes the propagation of light through the water column [34]. Legleiter et al. [3] estimated K_d by directly measuring a vertical profile of downwelling irradiance within a water column using a spectroradiometer with waterproof accessories. In this study, we solved for K_d using water-leaving reflectances observed for different known depths and a homogeneous bottom type adapted from [37–39]. For a small reach of river with a homogeneous bottom type, differences in bottom reflectance can be assumed negligible for a given pair of pixels, i.e., $r_{rs}^{B1} = r_{rs}^{B2}$. K_d can then be estimated by rearranging Equation (2) for each pair of pixels with different water depths (d_1, d_2):

$$K_d(\lambda) = \frac{ln\left(\frac{r_{rs2}(\lambda) - R_{rs}^{\infty}(\lambda)}{r_{rs1}(\lambda) - R_{rs}^{\infty}(\lambda)}\right)}{2(d_1 - d_2)}. \tag{3}$$

This approach for estimating K_d requires R_{rs} coupled with corresponding depth information. Water depth can be measured in the field or inferred from the image. Note that the depth samples for the estimation of K_d should be selected from a reach with a uniform substrate. However, the water depth of each individual pixel is required to estimate r_{rs}^{B} (Equation (2)), which can be retrieved from an image using bathymetry models such as Lyzenga's model [28,29], a band ratio model [40], optimal band ratio analysis (OBRA) [41,42], or multiple optimal depth predictors analysis (MODPA) [13].

We estimated water depths using MODPA, previously developed in the Sarca River, a method that has been proven to provide robust bathymetry retrievals with respect to substrate variability and water column heterogeneity [13]. MODPA initially increases the spectral domain of the original image by adding intensity components from RGB to the hue–saturation–intensity (HSI) transformations. All possible Lyzenga (Equation (4)) and ratio (Equation (5)) predictors of the produced high-dimensional image are then considered as candidate variables for a multiple regression bathymetry model. MODPA selects optimal predictors ($X^{Lyzenga\_Opt}$, X^{Ratio_Opt}) among all the candidates based on a feature selection method such as partial least square (PLS) regression to form the bathymetric model (Equation (6)). The unknown parameters of the model (a_i, b) can be estimated by performing a multiple regression between m optimal predictors and in-situ depths (d). Note that reflectances can be replaced with radiances in Equations (4) and (5) [12,20] for which R_{rs} was utilized in this study.

$$X^{Lyzenga} = ln(L_T(\lambda) - L_\infty(\lambda)), \quad L_\infty(\lambda) = L_C(\lambda) + L_S(\lambda) + L_P(\lambda) \tag{4}$$

$$X^{Ratio} = ln\frac{L_T(\lambda_1)}{L_T(\lambda_2)} \tag{5}$$

$$d = \sum_{i=1}^{m} a_i X_i + b, \quad X \in \left\{X^{Lyzenga_{Opt}}, X^{Ratio_{Opt}}\right\}. \tag{6}$$

The radiance observed over optically deep waters (L_∞) encompasses upwelling radiance from the water column (L_C), water surface (L_S), and atmosphere (L_P). Subtraction of L_∞ from the TOA radiance (L_T), known as a deep-water correction, isolates the radiance component upwelling from the bottom and can provide information about depth and substrate properties [29,42]. Correctly applying a deep-water correction is challenging in fluvial systems due to the lack of optically deep pixels. However, the effects of deep-water correction become important when the total radiance signal approaches the deep-water signal (i.e., the bottom-reflected signal becomes negligible). In shallow and clear rivers where bottom reflectance makes a larger contribution to the TOA radiance, deep-water correction has been dispensed in some applications such as bathymetric mapping [13,38,43,44]. However, the remote sensing reflectance of optically deep water (R_{rs}^{∞}) is also required to estimate K_d and bottom reflectance (Equations (2) and (3)). Legleiter et al. [3] collected spectra from the deepest part of the channel (~2 m deep) to obtain an estimate R_{rs}^{∞} for performing a water-column correction. However, this assumption is subject to significant uncertainties in clear or very shallow streams where bottom-reflected radiances are dominant. Flener [44] proposed an iterative procedure to estimate L_∞ or R_{rs}^∞ in shallow rivers that lack optically deep water: L_∞ or R_{rs}^∞ can initially be estimated using a first-guess and modified in an iterative process such that the correlation between image/spectra-derived quantities (X) and the water depths (d) is maximized. This research has utilized Flener's method [44] for estimating R_{rs}^∞ to assess its impact on retrieval of K_d and bottom reflectance.

Figure 2 illustrates the overall workflow for mapping bottom types without and with water-column correction, i.e., using (A) above-water reflectances and (B) retrieved bottom reflectances. The bottom information extracted from these two approaches is then compared to the reference data available from simulations or measured at the laboratory/field. The next subsection describes the methodology for classification of bottom classes and SAV densities.

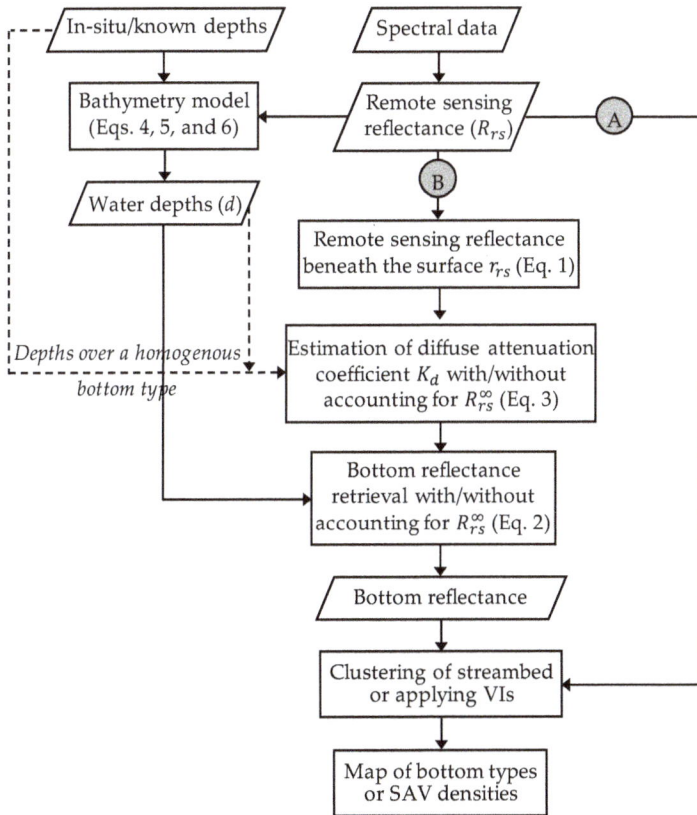

Figure 2. Flowchart for streambed mapping and delineation of SAV (A) before and (B) after water-column correction. The depth information required for K_d estimations can be collected either in the field or derived from image/spectra (shown by dashed lines).

3.2. Classification of Bottom-Type and SAV Densities

The application of supervised classification would be challenging in terms of collecting benthic samples for training the models. To broaden the applicability of our substrate mapping methodology, the k-means algorithm [45], a frequently used unsupervised classifier, was applied to both above-water and retrieved bottom reflectances to map riverbed clusters. The labels were then assigned to the clusters based on interpretation of the spectra associated with the clusters' centers.

Remote Sens. **2019**, *11*, 262

This research also investigates the effectiveness of widely used terrestrial and aquatic vegetation indices (VIs) for detecting and quantifying SAVs in shallow rivers. VIs with different band combinations were examined to identify the SAVs with different densities before and after water-column correction. WV3 is equipped with a red-edge (RE) band and has two NIR bands (NIR1 and NIR2), which collectively allow for evaluating more VIs compared to four-band imagery such as GeoEye data.

The normalized difference vegetation index (NDVI) is a commonly used index [46,47] for examining properties of terrestrial vegetation (Table 3). The attenuation by the water-column, however, can influence the water-leaving radiance to varying degrees depending on depth, bottom properties, and IOPs. More specifically, the sharp increase in reflectance within red to NIR transition spectrum becomes attenuated due to strong pure water absorption in the NIR region [48]. Recently, a water-adjusted vegetation index (WAVI) (see Table 3) has been suggested to account for the background water response [11]. However, this index was developed and tested only in a few lakes. In this study, alternative NDVI and WAVI were computed by replacing the traditional NIR band (~851 nm for GeoEye) with RE and NIR2 bands of WV3 (Table 3).

Table 3. Vegetation indices (VIs) used to study SAV.

VIs	Original Formula	Alternative WV3 Band Combinations
Terrestrial	$WAVI = \dfrac{R_{NIR} - R_R}{R_{NIR} + R_R}$	(NIR1, R), (NIR2, R), (RE, R)
Aquatic	$WAVI = 1.5\dfrac{R_{NIR} - R_B}{R_{NIR} + R_B + 0.5}$	(NIR1, B), (RE, B)

VIs such as the NDVI are widely used as indicators for fractional vegetation coverage [49,50]. To evaluate the effectiveness of VIs for quantifying SAV densities, using simulated data, regression analyses were performed for various VIs and SAV fractions. In addition, the clustering of VIs to distinguish among SAV density classes was evaluated using laboratory and WV3 data.

3.3. Accuracy Assessment

The root mean square errors (RMSEs) were calculated to assess the retrievals of bottom reflectance (R^B) and K_d ($RMSE\_R$ and $RMSE\_K_d$, respectively). Here, we assumed that the bottom is Lambertian for converting the r_{rs}^B to the unitless reflectance, i.e., $R^B = \pi \times r_{rs}^B$ [51]. Note that this assumption is subject to uncertainty due to probable non-Lambertian behavior of the riverbed. However, this does not affect the relative comparison of the spectra as R is a factor of r_{rs}^B.

$$RMSE\_R = \sqrt{\frac{\sum_{i=1}^{n}\left(R^{B,reference}(\lambda_i) - R^{B,retrieved}(\lambda_i)\right)^2}{n}} \qquad (7)$$

$$RMSE\_K_d = \sqrt{\frac{\sum_{i=1}^{n}\left(K_d^{reference}(\lambda_i) - K_d^{retrieved}(\lambda_i)\right)^2}{n}} \qquad (8)$$

where the "reference" superscript refers to known parameters from either simulations or measurements. The estimated parameters are denoted by a "retrieved" superscript. n is the number of bands for which visible ($\lambda < 700$ nm) and NIR ($\lambda > 700$ nm) bands were analyzed separately in this study. Note that $RMSE\_R$ is unitless, while $RMSE\_K_d$ has units of 1/m.

Statistics derived from a confusion matrix were also used for the assessment of substrate classes. For a classified map, overall accuracy is the number of correctly classified pixels divided by the total number of pixels. The kappa statistic is a measure of how the classification results compare to class allocations assigned by chance, which is a pessimistic estimation of accuracy. The producer accuracy provides a measure of accuracy for each individual class by calculating the fraction of correctly classified pixels of a given class with respect to the total number of reference pixels for the same class.

The user accuracy presents the reliability of each class and is calculated as the fraction of correctly classified pixels of a given class with respect to the total number of pixels labeled as the same class in the classified map [46].

4. Experiments and Results

The K_d and bottom reflectance retrieval methods were applied to the datasets described in Section 2 and the associated analyses and results are presented in the following subsections. An overall evaluation of the results is then provided in the discussion section. Note that, hereafter, the spectral parameters are distinguished for each experiment using Lab, Sim and WV3 superscripts for laboratory, simulated, and WV3 data, respectively (e.g., R_{rs}^{Sim} stands for the simulated R_{rs}).

4.1. Laboratory Radiometric Measurements

To quantify water-leaving reflectance under various conditions, including depth and bottom type, spectroscopic measurements were performed in a flume at the University of Trento's hydraulic laboratory. Spectral reflectance measurements were acquired in a darkroom with an Analytical Spectral Devices (ASD) HandHeld 2 spectroradiometer operating within the 325–1075 nm spectral range. A standard ASD illuminator was used to produce a highly stable light source across the full visible/NIR spectral range. The spectral data were recorded by pointing a fiber optic jumper cable in a near-nadir viewing angle 30 cm above the water surface. The sensor's field of view was adjusted to sample a cell in the center of the channel to avoid any adjacency effects associated with the flume sidewalls. The illumination geometry was modified to eliminate instrument self-shading over the flume [52]. Three spectra were recorded for each flow condition by averaging 25 individual samples. Radiometric calibrations including white reference and dark current observations were updated before each set of measurements to collect data in reflectance mode.

Four sets of data were collected over different bottom types, including a non-vegetated gravel bed and three SAV densities (high, medium, and low). For each set, dry bottom reflectance (representing exposed material) was first measured as the reference bottom reflectance. Measurements were then continued with 1 cm increments in the water level up to 40 cm. Figure 3 shows the hydraulic flume and the configuration of spectroscopic measurements.

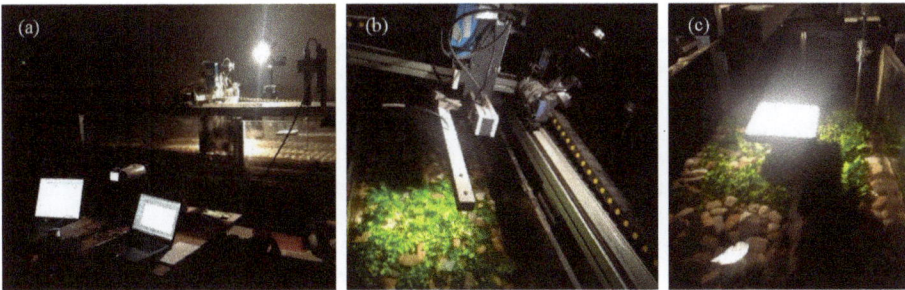

Figure 3. (**a**) The setup for spectroscopic experiments on the hydraulic flume and the spectral measurements of (**b**) SAV with high density and (**c**) the white reference. The spectral reflectances were recorded from above the water surface within the 325–1075 nm range using a fiber optic cable connected to an ASD HandHeld 2 spectroradiometer.

The reflectances (R^{Lab}) collected over the water surface were converted to R_{rs}^{Lab} and convolved with GeoEye ($R_{rs}^{Lab,G}$) and WV3 ($R_{rs}^{Lab,WV3}$) spectral band passes. R_{rs}^{Lab} along with the bathymetry data collected over a non-vegetated gravel bed were then used to estimate K_d (Equation (3)). r_{rs}^B was then retrieved using Equation (2).

The above-water reflectances are shown with the associated retrieved bottom reflectances in Figure 4a, allowing for a comparison of 40-cm-deep water at low, medium, and high densities of SAV. The characteristic feature of vegetation is evident on the retrievals of bottom reflectances. However, this feature is significantly attenuated for above-water spectra such that the feature completely disappears for the low-density SAV. The retrieved bottom reflectance showed good agreement (Equations (7) and (8)) with the measured reflectances, particularly across the visible spectrum, and the deep-water correction slightly improved the results (Figure 4b,c). The retrievals from $R_{rs}^{Lab,G}$ (Figure 4c) led to slightly lower RMSEs over the NIR spectrum compared to those from $R_{rs}^{Lab,WV3}$ (Figure 4b). This is because WV3 includes an additional band (NIR2; Table 1) spanning over longer NIR wavelengths where pure water absorption is much stronger. The error bars in Figure 4 indicate the effects of the changing water level, i.e., the smaller the error bars, the better the water-column correction. Note that RMSEs have been estimated using five visible bands ($\lambda < 700$ nm) and three NIR bands ($\lambda > 700$ nm) using $R_{rs}^{Lab,WV3}$. The RMSEs of reflectance retrievals (i.e., $RMSE\_R$) for high-density SAV were slightly higher (~0.01) in the NIR spectrum, particularly when the deep-water correction was not applied. This can be attributed to strong attenuation of the vegetation feature in the NIR region caused by pure water absorption. However, deep-water correction mitigated this effect.

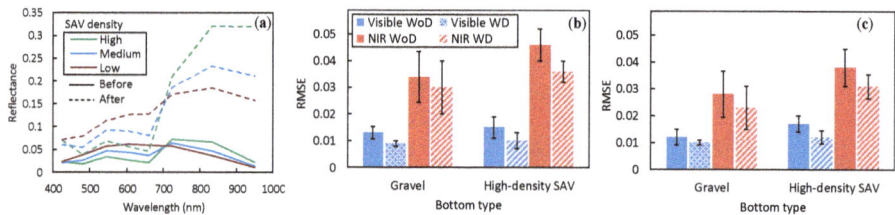

Figure 4. (**a**) Spectra characterized by different SAV densities before and after water-column correction ($R_{rs}^{Lab,WV3}$ vs. $r_{rs}^{B,WV3}$). RMSEs for the retrieved bottom reflectances (unitless) with (WD) and without (WoD) applying deep-water correction across visible and NIR bands (Equation (7)) using (**b**) $R_{rs}^{Lab,WV3}$ and (**c**) $R_{rs}^{Lab,G}$. Error bars (standard deviation of $RMSE\_R$) indicate the effect of variable water depth.

Using either above-water spectral measurements R_{rs}^{Lab} or retrieved bottom reflectances $r_{rs}^{B,Lab}$ for various SAV densities, original and alternative NDVIs and WAVIs (Table 3) were computed. The VIs derived from each band combination led to thematic clusters associated with the four SAV densities (see Figure 5; each SAV density is shown with a different color). The thematic clusters of VIs derived from R_{rs}^{Lab} show considerable overlap, which reduces separability among different SAV densities. To further elaborate, the k-means algorithm was applied to VIs to automatically cluster them into four classes. The clusters were ranked based on the average magnitude of the calculated VIs and accordingly were assigned to SAV density classes (i.e., the higher the VI magnitude, the higher the SAV density). The overall accuracies and the kappa coefficients are presented for VIs with different band combinations (Figure 5). The VIs built upon the RE band demonstrated better performance compared to other band combinations using R_{rs}^{Lab}. More specifically, the (RE, R) band combination yielded the highest accuracy with 92% overall accuracy and kappa coefficient of 89%. The aquatic VIs provided no further benefit for clustering SAV densities using R_{rs}^{Lab}. The clusters obtained from $r_{rs}^{B,Lab}$ indicated remarkable distinctions among SAV densities for all the band combinations (Figure 5b). In addition, the clusters are very compact, suggesting minimal impact of varying water depth; this result confirms successful correction of water-column effects. However, these results are based on observations which were limited to a maximum depth of 40 cm with minimal constituent loads (see Table 3). The results based on water-column correction (Figure 5b) are shown only for the case without applying the deep-water correction, as no more enhancements were required for clustering the SAVs.

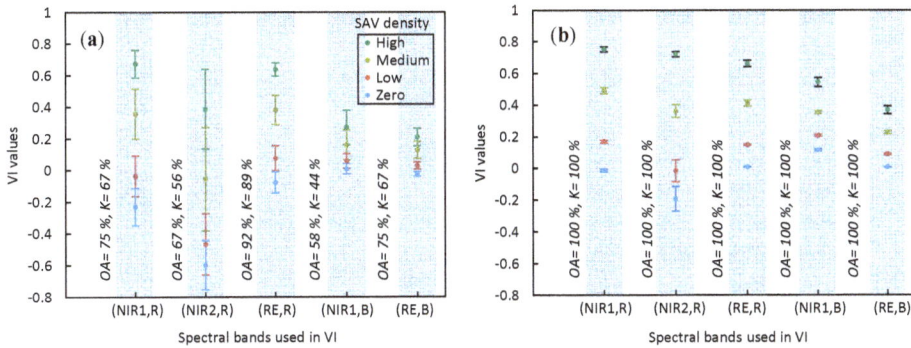

Figure 5. Evaluating the effectiveness of VIs in clustering SAVs with different densities in laboratory experiments. k-means clustering is applied on VIs with different band combinations (**a**) before and (**b**) after water-column correction. Zero-density SAVs stand for non-vegetated gravel bed. OA: overall accuracy; K: kappa coefficient.

4.2. Radiative Transfer Simulations

Simulated spectra produced by radiative transfer modeling have been used previously for studying bathymetry retrieval in shallow rivers [12,13,42,53]. Building on this approach, we utilized simulated spectra to gain insight into streambed mapping in shallow riverine environments. To investigate the effectiveness of water-column correction for varying IOPs and bottom types, we simulated the optical properties (Table 2) and generated substrate spectral mixtures for a reach of the Sarca River. The R_{rs} as well as the associated K_d were simulated with the widely used Hydrolight radiative transfer model [26,51] for three different bottom types (macrophyte, dark sediment, and dolomite) as well as a range of water column constituents representative of the Sarca River and similar alpine rivers. Maximum and minimum values of the constituents were selected based on long-term observations of water quality indicators documented by local environmental agencies [25]. A database of simulations was produced including more than 20,000 individual spectra (Table 2).

The bathymetry of a reach of the Sarca River was derived from the WV3 image using MODPA [13] and used as a basis for the simulations (Figure 6a). Only a randomly selected 1% of the entire channel depths (about 50 pixels) were used to calibrate the MODPA model and the remaining known depths were reserved for validating the bathymetry model (Figure 6b). The resultant coefficient of determination (R^2) of 0.99 and an RMSE of 0.01 m indicated the robustness of the depth retrieval method with respect to the variability in constituents and substrate types within the channel (Figure 6).

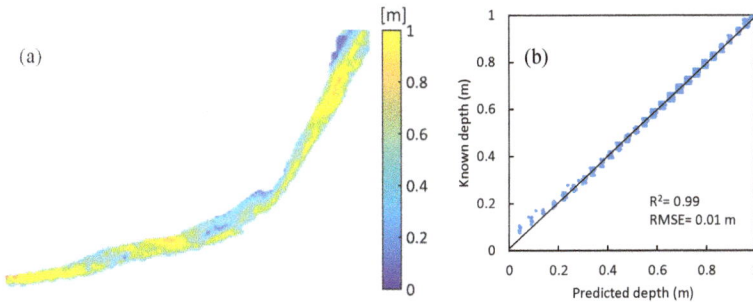

Figure 6. (**a**) The bathymetry map retrieved for the simulated channel using $R_{rs}^{Sim\_Channel,WV3}$ and (**b**) match-up validation.

The channel was divided into three segments with different concentrations of constituents from clear to turbid water (constituents associated with each of the segments are shown in Figure 7c). Each segment has one dominant bottom type (Figure 7b) but is mixed with up to 50% of two other bottom types. Note that R_{rs} spectra were first simulated considering pure bottom types. The spectra for constant/known water depths (d) and constituents, and only with different bottom types (i.e., $R_{rs}^{B_1}$, $R_{rs}^{B_2}$ and $R_{rs}^{B_3}$), were then mixed linearly [11,54] with the desired fractions (f_1, f_2, f_3) to produce the R_{rs} for the simulated channel ($R_{rs}^{Sim\_Channel}$) according to Equation (9). To generate a reference map for accuracy assessment, each pixel was labeled as the bottom type with the largest fraction (Figure 7b). Figure 7 shows the inputs for simulating the spectra over the river channel. Note that the arrangements of bottom types and constituents within the reaches are just to allocate the simulated spectra to individual pixels so that each reach has a specific optical condition. However, the analyses were performed at the pixel level and independent from the spatial distribution of the pixels.

$$R_{rs}^{Sim\_Channel} = f_1 \times R_{rs}^{B_1} + f_2 \times R_{rs}^{B_2} + f_3 \times R_{rs}^{B_3}, \; f_1 + f_2 + f_3 = 1 \quad (9)$$

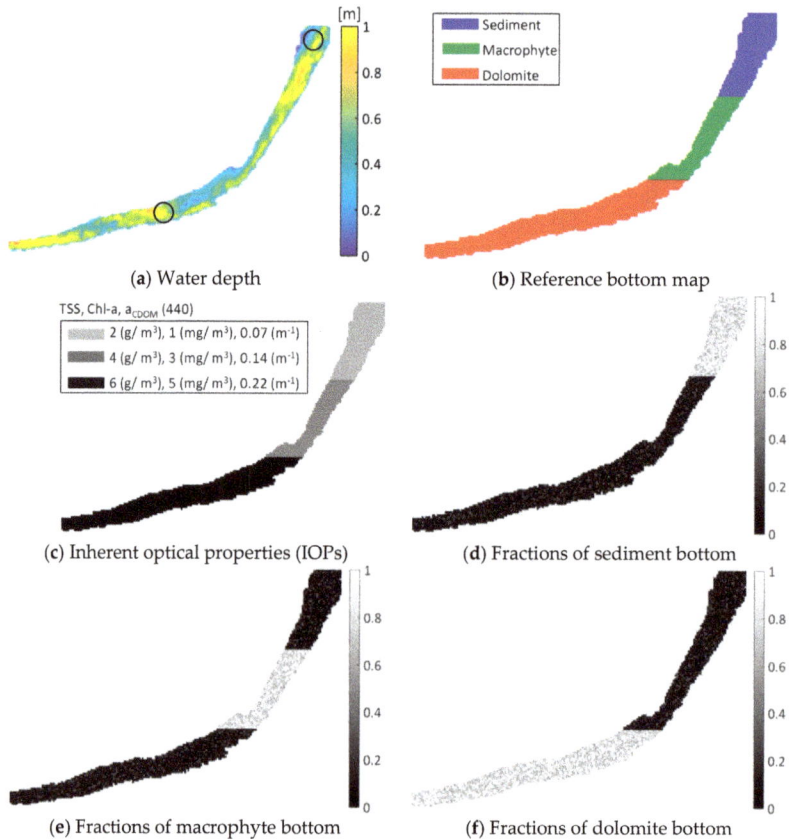

(a) Water depth

(b) Reference bottom map

(c) Inherent optical properties (IOPs)

(d) Fractions of sediment bottom

(e) Fractions of macrophyte bottom

(f) Fractions of dolomite bottom

Figure 7. Inputs for simulation of spectra across the river channel ($R_{rs}^{Sim\_Channel}$) associated with (**a**) water depths in a reach of the Sarca River considering (**b**) dominant bottom types, (**c**) variable constituents, and (**d–f**) fractions of bottom types. A linear mixture model (Equation (9)) was applied to account for the variability (fractions) of bottom types within the pixels. Two test sites used for the estimation of K_d are highlighted by circles on the first graph.

Furthermore, we examined the performance of K_d retrieval in various water-column conditions in the simulated river channel with $R_{rs}^{Sim\_Channel}$ as the reference. This analysis allowed us to evaluate the effects of variable constituents on the retrievals of K_d and r_{rs}^B. As required in Equation (3), a number of samples (about 20 pixels) of water depths and the associated $R_{rs}^{Sim\_Channel}$ were taken from two different test sites: (a) upstream composed of a dominant substrate type of sediment and with a relatively clear water column and (b) downstream with a dominant substrate type of dolomite and relatively turbid waters (Figure 7). The estimations of K_d with and without accounting for R_{rs}^∞ [44] were compared with the average $K_d^{Sim\_Channel,WV3}$ within the entire channel as a reference (Figure 8).

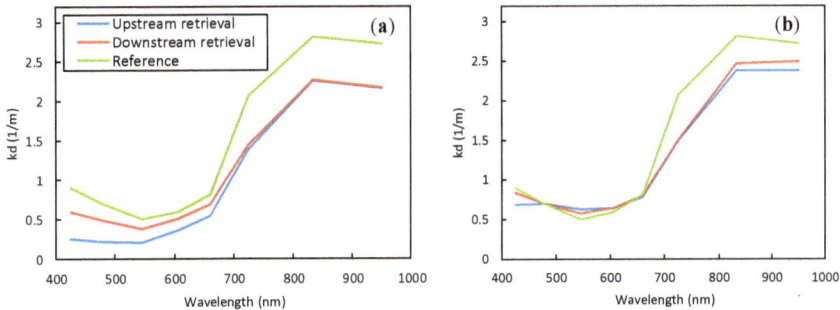

Figure 8. Retrievals of K_d for the simulated river channel (**a**) without and (**b**) with accounting for R_{rs}^∞ compared to the reference $K_d^{Sim\_Channel,WV3}$.

Note that constituents assigned to the downstream area were more representative for the entire simulated stream. Therefore, as expected, downstream estimates of K_d were more in agreement with the reference $K_d^{Sim\_Channel}$ (Figure 8) than that of the upstream retrievals, particularly when R_{rs}^∞ was not taken into account. Figure 9 shows the RMSEs for upstream and downstream K_d retrievals with and without applying deep-water correction. Considering R_{rs}^∞ led to improvements in deriving K_d, particularly for the upstream-based retrieval.

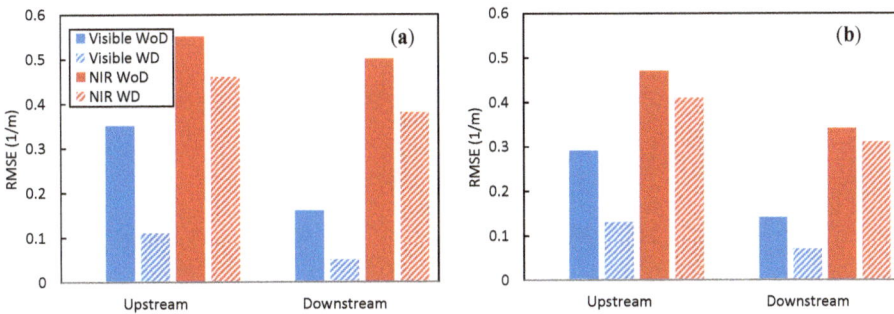

Figure 9. Performance of K_d retrievals expressed in terms of RMSE (1/m) using (**a**) $R_{rs}^{Sim\_Channel,WV3}$ and (**b**) $R_{rs}^{Sim\_Channel,G}$ by sampling the depths from upstream and downstream of the simulated channel with (WD) and without (WoD) accounting for deep-water reflectance R_{rs}^∞.

We utilized the k-means algorithm to perform riverbed classification using $R_{rs}^{Sim\_Channel}$ and $r_{rs}^{B,Sim\_Channel}$. As evident in Figure 10, above-water reflectances ($R_{rs}^{Sim\_Channel}$) led to a considerable number of misclassified pixels, particularly confusion between the bottom types of the upstream (dominant sediment) and downstream (dominant dolomite) segments. Substrate clusters derived after water-column correction (i.e., using $r_{rs}^{B,Sim\_Channel}$) showed considerably fewer misclassified pixels.

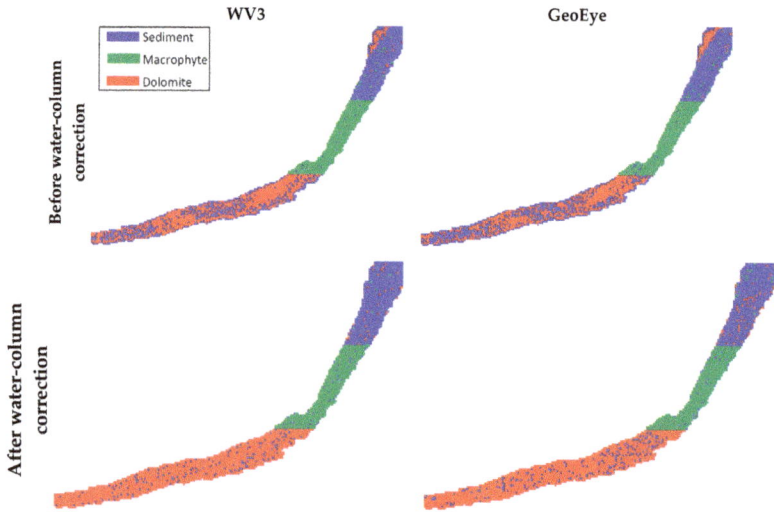

Figure 10. Clustering of bed-types for simulated channel before and after water-column correction (i.e., using $R_{rs}^{Sim\_Channel}$ and $r_{rs}^{B,Sim\_Channel}$, respectively) by sampling the pixels required for K_d retrieval from upstream.

The water-column correction yielded significant improvements (about 20% overall accuracy and 30% kappa coefficient) in mapping bottom types using $R_{rs}^{Sim\_Channel}$ either for WV3 or GeoEye spectra (Figure 11). Accounting for R_{rs}^{∞} slightly improved the bottom mapping (~2–3%). Further, downstream K_d retrievals yielded slightly better results compared to those of upstream (~2–3%), which implies that substrate mapping was independent of constituent variability in our case study. However, this would probably have considerable effects when detailed spectral information is required for mapping the substrate properties (e.g., bottom types with very similar spectral properties such as different types of SAV) or in the case of highly variable constituents.

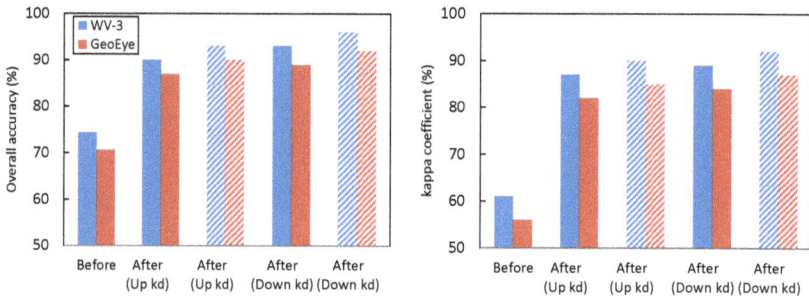

Figure 11. The overall accuracies and kappa coefficients of the bottom maps before and after water-column correction. The statistics are presented for upstream (Up) and downstream (Down) K_d retrievals for WV3 and GeoEye spectra. The accuracies with applying deep-water correction are shown by hatched bars.

To evaluate the effectiveness of VIs for the detection of SAV densities, a regression analysis was performed between known SAV (macrophyte) fractions and associated VIs. The R^2 and RMSE for this analysis indicated a significantly stronger correlation between VI and SAV densities after water-column correction (Figure 12). This finding was valid for all band combinations. The strongest correlation was for (RE, R) band combination (Table 1) using either $R_{rs}^{Sim\_Channel,WV3}$ ($R^2 = 0.48$ and RMSE = 0.2) or inferred $r_{rs}^{B,Sim\_Channel,WV3}$ ($R^2 = 0.85$ and RMSE = 0.07). This demonstrated the significance of the

WV3's RE band (i.e., 742 nm) for mapping benthic vegetation in shallow streams, but this band is not available on GeoEye. There is also evidence from studies in wetlands demonstrating the usefulness of the RE band for mapping benthic vegetation but without applying any water-column correction [55,56]. Accounting for R_{rs}^∞ slightly improved the regression statistics.

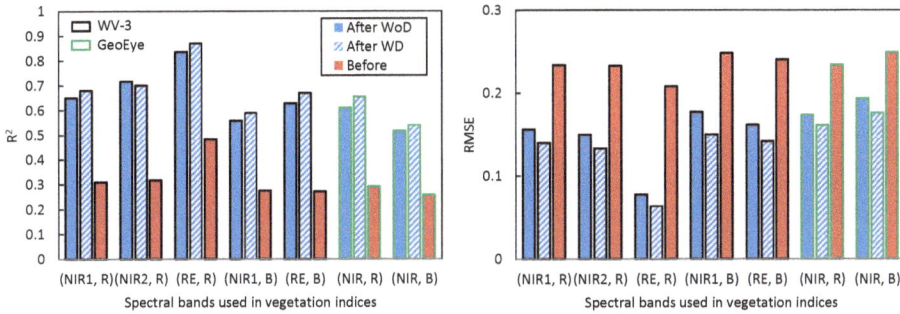

Figure 12. R^2 and RMSE of regressions between VIs and SAV fractions before and after water-column correction with (WD) and without (WoD) accounting for deep-water reflectance R_{rs}^∞.

Additional analyses were performed on the database of simulated spectra to examine the effects of water depth and constituents on the bottom reflectance retrieval. Figure 13 (first row) indicates the RMSEs for the inferred bottom reflectances across a range of water depths while holding the constituents constant (TSS = 4 g/m^3, Chl-a = 3 mg/m^3, and a_{CDOM} (440) = 0.14 m^{-1}). In general, reflectances within the visible bands were retrieved with high accuracies, and the water depth had little effect on RMSEs. The RMSEs for the NIR bands increased sharply with water depth particularly without correcting for R_{rs}^∞. The effect of accounting for R_{rs}^∞ was pronounced for relatively deep waters (depth > 0.5 m) and improved the bottom reflectance retrievals, particularly in the NIR spectrum.

In addition, some analyses were performed to investigate the effect of variations in constituent concentration on bottom reflectance retrievals. Three realistic levels of turbidity assumed for the range of constituents in the Sarca River and similar Alpine rivers were considered: low (TSS = 2 g/m^3, Chl-a = 1 mg/m^3, a_{CDOM} (440) = 0.07 m^{-1}), medium (TSS = 4 g/m^3, Chl-a = 3 mg/m^3, a_{CDOM} (440) = 0.14 m^{-1}), and high (TSS = 6 g/m^3, Chl-a = 5 mg/m^3, a_{CDOM} (440) = 0.22 m^{-1}). These constituent conditions are labeled as low, medium, and high turbidity in Figure 13. The effects of constituents were then evaluated in a constant and relatively thick water column (1 m). The RMSE for bottom reflectance retrievals in the NIR bands increased with the increase in turbidity, while the retrievals in the visible bands were less affected. Accounting for R_{rs}^∞ improved the retrieval of bottom reflectance, particularly in the NIR bands (see the second row in Figure 13).

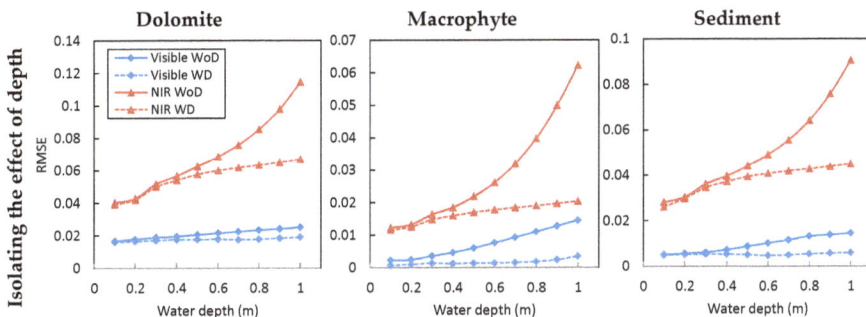

Figure 13. *Cont.*

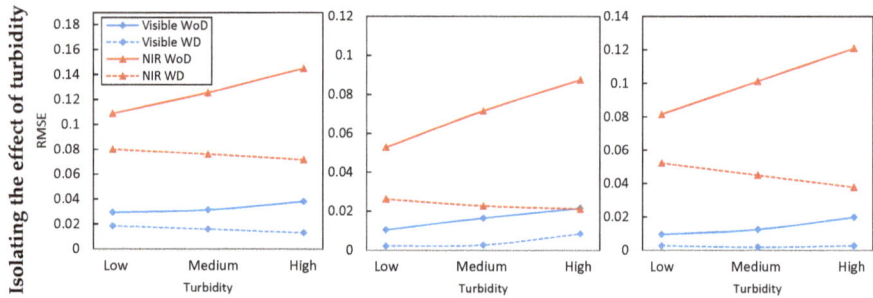

Figure 13. RMSEs (unitless) of reflectance retrievals using the database of $R_{rs}^{Sim\_WV3}$ for three bottom types by isolating the effect of water depth (constant constituents: TSS = 4 g/m^3, Chl-a = 3 mg/m^3, a$_{CDOM}$ (440) = 0.14 m^{-1}) and the effect of constituents (constant 1 m deep water) with (WD) and without (WoD) accounting for deep-water reflectance R_{rs}^∞.

4.3. Image Analysis and Field Survey

To gauge the performance of the bottom reflectance retrieval methodology for SAV-density mapping, we examined an eight-band WV3 image (Table 1) of the Sarca River. The image was acquired on 1 September 2015 with a mean off-nadir view angle of 10.5° and a 1.6 m spatial resolution. In-situ water depths and information on SAV densities were recorded using a real-time-kinematic (RTK) GPS rover (Figure 14). The in-situ depth measurements were conducted along cross sections in three reaches only a few days after image acquisition. Note that the river is regulated by a dam upstream and the water level remained highly stable. The image was delivered georeferenced, but we used control points collected outside the river channel to improve for accurate co-registration of the in-situ data with the image. To link field depths to image pixels, an ordinary kriging was used to interpolate the measured depths at each pixel location [20]. One-half of the data was used for calibration of the MODPA model and the second half as validation for accuracy assessment. For each patch of SAV, approximate areal coverage was documented to further evaluate the performance of clustering SAV-density classes.

Figure 14. (a) Field observations to record water depths and SAV densities assisted by a precise RTK GPS, (b) cross-sectional measurement points of water-depths superimposed on the WV3 image, and (c) a sample of SAV in the Sarca River. The sampling area used for estimation of K_d is highlighted by a circle on the satellite image.

A near-simultaneous Landsat-8 image was processed via the SeaWiFS Data Analysis System (SeaDAS) to infer dominant aerosol models over Lake Garda. The eight-band WV3 image of the study area was then atmospherically corrected using the MODerate resolution atmospheric TRANsmission (MODTRAN) code [57,58] to provide R_{rs}^{WV3}. K_d^{WV3} was then estimated using in-situ depths and associated R_{rs}^{WV3} (Equation (3)) over a segment of the Sarca River with homogeneous bottom type (shown on Figure 14b). The water-depth map was estimated by calibrating the MODPA model by randomly selecting half of the in-situ depths. The image-derived depth map is shown in Figure 15b for a subset of the river where field observations were carried out (Figure 15a). The validation was performed using the remaining half of the in-situ depths (Figure 15c).

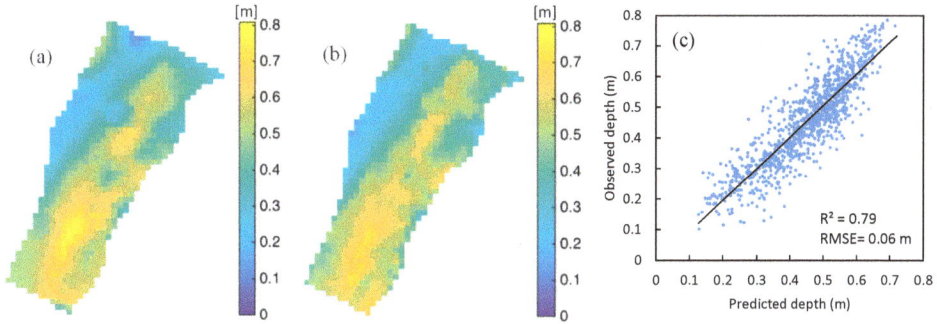

Figure 15. (**a**) In-situ depths compared to (**b**) MODPA-derived depth map of the Sarca River and (**c**) match-up validation.

The areal coverage of SAV patches gathered in the field was converted to a density index by dividing the observed area of a patch by the spatial resolution of the image (1.6 × 1.6 m). As a reference map, the index values were clustered using k-means algorithm to three density classes (Figure 16a). The image-derived VIs (either before or after water-column correction) were also clustered using the k-means algorithm and then compared with the reference map. The best results were achieved when the (RE, R) band combination was used both with or without applying water-column correction (Figure 16b,c). These findings are consistent with the results obtained from laboratory and synthetic data analyses.

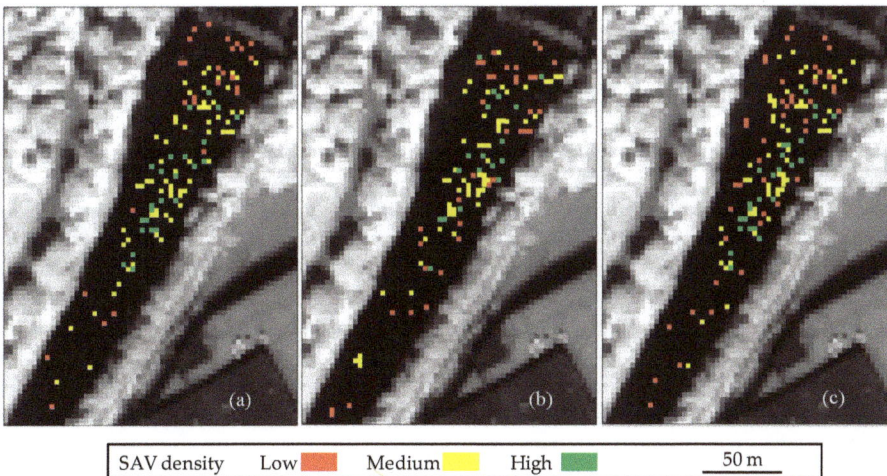

Figure 16. (**a**) In-situ map of SAV densities compared to the maps derived from clustering of the VI with (RE, R) band combination (**b**) before and (**c**) after water-column correction using the WV3 image.

The user and producer accuracies of SAV-density clusters indicated that the retrieved $r_{rs}^{B,WV3}$ yielded remarkably higher accuracies than R_{rs}^{WV3} for all the SAV densities. These enhancements were on the order of 22% and 34% in average user and producer accuracies, respectively. Note that accounting for R_{rs}^{∞} has improved the average user/producer accuracies on the order of 5% for the SAV-density mapping based on $r_{rs}^{B,WV3}$. The accuracies of clustering from R_{rs}^{WV3} improved by increasing the SAV density (45% user accuracy and 57% producer accuracy for high-density SAV). This is also valid for clustering from $r_{rs}^{B,WV3}$ with a lower magnitude (Figure 17).

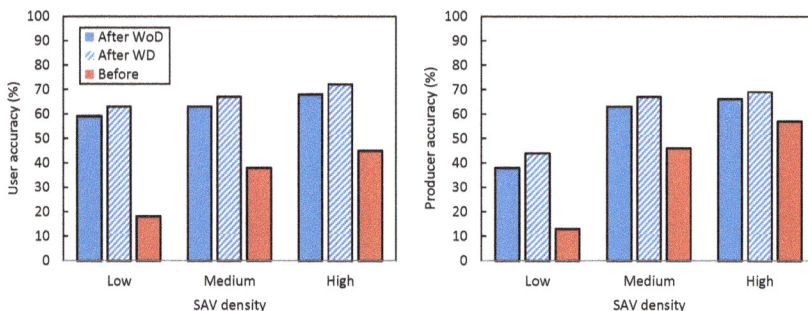

Figure 17. User and producer accuracies of SAV density clusters derived from WV3 image based on VI (RE, R) before and after water-column correction with (WD) and without (WoD) accounting for deep-water reflectance, R_{rs}^{∞}.

5. Discussion

Bottom reflectance retrieval and substrate-type mapping were explored via three independent experiments. The spectroscopic measurements in the hydraulic laboratory and simulations from radiative transfer modeling provided a thorough understanding of the driving factors influencing the feasibility and accuracy of streambed mapping, such as water depth, constituents, deep-water correction, and choices of spectral bands. Further, in a first attempt to map substrate properties from space, an eight-band WV3 image of a reach in the Sarca River was used to classify SAV densities. Results based on spectroscopic measurements and simulations suggest that K_d and bottom reflectance retrieval was more accurate in the visible bands than in the NIR bands, particularly for relatively deep waters (>0.5 m). This is attributed to the rapid light attenuation towards longer wavelengths in the NIR region, particularly for thicker water columns. Accounting for deep-water reflectance, R_{rs}^{∞} was demonstrated to be effective for enhancing retrievals in the NIR spectrum when the water becomes deeper. This result is reasonable, as applying R_{rs}^{∞} has more of an effect when the bottom-reflected component of the water-leaving radiance approaches zero. However, the effect of R_{rs}^{∞} was negligible for visible bands in the range of water depths discussed in this study (<1 m) as well as for the NIR bands in very shallow depths (<0.5 m). Further analyses using synthetic data revealed that IOP variability has less impact on $r_{rs}^{B,Sim}$ (bottom reflectance) retrieval in the visible bands. Increasing turbidity reduced the accuracy of $r_{rs}^{B,Sim}$ retrieval in the NIR bands. However, accounting for R_{rs}^{∞} mitigated the effect of turbidity on retrieval of $r_{rs}^{B,Sim}$. For instance, *RMSE_R* for the macrophyte bottom was reduced ~4X when applying the deep-water correction in highly turbid waters (Figure 13).

As a key finding, water-column correction significantly improved riverbed mapping. For instance, retrieval of $r_{rs}^{B,Sim\_Channel}$ for three bottom types (dolomite, macrophyte, and sediment) within the simulated channel (Section 4.2) enhanced the riverbed clustering on the order of 20% in overall accuracy and 30% in kappa coefficient compared to classifications obtained from $R_{rs}^{Sim\_Channel}$. This was also demonstrated in distinguishing among SAV densities where retrieval of bottom reflectance yielded VIs strongly correlated with macrophyte fractions. The terrestrial VI with (RE, R) band combination was found to provide the highest correlation with the SAV fractions using either $r_{rs}^{B,Sim}$ ($R^2 = 0.85$ and RMSE = 0.07) or R_{rs}^{Sim} ($R^2 = 0.48$ and RMSE = 0.2). The same band combination also yielded the most accurate clusters of SAV densities in analyzing laboratory data as well as the WV3 image. The above-water reflectances (R_{rs}^{WV3}) showed potentials for detecting high-density SAVs in the Sarca River (user accuracy = 45% and producer accuracy = 57%). This indicates the effectiveness of WV3's RE band (i.e., 724 nm) for mapping SAVs. Moreover, enhanced spectral resolution of the WV3 compared to the GeoEye provided higher accuracies (on the order of 5%) in mapping the streambed using synthetic data. Note that K_d retrievals for the NIR2 band was slightly less accurate than that derived for GeoEye's NIR band. This can be attributed to the strong water-column attention in the NIR2. Nevertheless, the improved accuracies gained in the clustering experiment using a WV3 image indicated the overall efficacy of enhanced spectral resolution of this sensor compared to the traditional four-band high resolution satellite imagery for mapping bottom compositions. We performed the analyses independent from the spatial resolution in order to isolate the effect of the spectral resolution of sensors. The spatial resolution can also affect the retrieval of bottom reflectances, particularly when there is a high level of mixture in the bed type. However, the effect of spatial resolution would be minor in our case due to comparable spatial resolutions of WV3 and GeoEye imagery.

6. Conclusions and Outlook

Recent research has generated significant optimism regarding the potential of optical remote-sensing imagery to extract key hydro-morphological attributes (e.g. bathymetry) of riverine environments. Understanding and isolating the effect of individual river attributes on the overall spectral response of a water body would reveal valuable information and could enable a wide range of applications in fluvial systems. Although studies of river bathymetry have become relatively mature, little work has been done to explore other essential attributes such as streambed composition. In this research, retrieval of bottom reflectances and mapping riverbed types were addressed. Unlike the bulk of the existing literature [10,17,48], a water-column correction approach was pursued to map bottom properties by retrieving bottom reflectance rather than using above-water spectra. This methodology accounted for water-column attenuation by estimating K_d using known depths with a homogeneous bottom type. MODPA was implemented to empirically derive the bathymetry and provided robust depth retrieval. Image-derived depths were then used for estimating K_d and then bottom reflectance, so that no in-situ optical measurements were required to obtain K_d.

Our attempt to retrieve bottom reflectance from space using WV3 image data, with a focus on mapping SAV densities, demonstrated promising results in a shallow riverine environment. However, further studies are needed to investigate mapping various benthic covers and other substrate attributes (e.g., grain sizes). Sun glint can be a source of uncertainty for mapping bottom types and compositions [59,60]. Imagery affected by sun glint would require pre-processing to reduce the undesirable surface reflections. K_d and bottom reflectance retrieval was also facilitated by bathymetric information, which requires some in-situ depth measurements. However, this approach undermines the full potential of streambed mapping when in-situ depth observations are lacking. Theoretical calibration methods of bathymetry models, such as the hydraulically assisted bathymetry model [61], can overcome this problem. Therefore, integration of streambed mapping methodologies with bathymetry models built upon theoretical calibration should be addressed in future studies. The effectiveness of pan-sharpening methods can also be examined in future works in order to further enhance the spatial resolution of streambed mapping using WV3 imagery. Moreover, applications of publicly free Sentinel-2 imagery would be interesting for mapping bottom compositions in large rivers with wider reaches.

Author Contributions: Conceptualization, M.N.J., N.P., and A.V.; Data curation, M.N.J., N.P., and A.V.; Formal analysis, M.N.J.; Methodology, M.N.J. and N.P.; Project administration, M.N.J. and A.V.; Resources, M.N.J. and A.V.; Software, M.N.J.; Validation, M.N.J.; Visualization, M.N.J.; Writing—original draft, M.N.J.; Writing—review & editing, M.N.J., N.P., and A.V.

Funding: Nima Pahlevan was funded under NASA ROSES #NNX16AI16G and the USGS Landsat Science Team Award #140G0118C0011.

Acknowledgments: This work was carried out within the SMART Joint Doctorate (Science for the MAnagement of Rivers and their Tidal systems) funded with the support of the Erasmus Mundus programme of the European Union. The authors gratefully acknowledge ASD Inc. for awarding a short-term usage of HandHeld2 spectroradiometer as of Alexander Goetz Instrument Support program. We are also grateful to Prof. Aronne Armanini, Prof. Maurizio Righetti, and Prof. Guido Zolezzi for giving us access to the hydraulic laboratory.

Conflicts of Interest: The authors declare no conflict of interest.

References

1. Westaway, R.M.; Lane, S.N.; Hicks, D.M. Remote sensing of clear-water, shallow, gravel-bed rivers using digital photogrammetry. *Photogramm. Eng. Remote Sens.* **2001**, *67*, 1271–1281.
2. Woodget, A.S.; Carbonneau, P.E.; Visser, F.; Maddock, I.P. Quantifying submerged fluvial topography using hyperspatial resolution UAS imagery and structure from motion photogrammetry. *Earth Surf. Process. Landf.* **2015**, *40*, 47–64. [CrossRef]
3. Legleiter, C.J.; Stegman, T.K.; Overstreet, B.T. Spectrally based mapping of riverbed composition. *Geomorphology* **2016**, *264*, 61–79. [CrossRef]
4. Chen, Q.; Yu, R.; Hao, Y.; Wu, L.; Zhang, W.; Zhang, Q.; Bu, X. A New Method for Mapping Aquatic Vegetation Especially Underwater Vegetation in Lake Ulansuhai Using GF-1 Satellite Data. *Remote Sens.* **2018**, *10*, 1279. [CrossRef]
5. Lane, S.N.; Widdison, P.E.; Thomas, R.E.; Ashworth, P.J.; Best, J.L.; Lunt, I.A.; Sambrook Smith, G.H.; Simpson, C.J. Quantification of braided river channel change using archival digital image analysis. *Earth Surf. Process. Landf.* **2010**, *35*, 971–985. [CrossRef]
6. Strayer, D.L. Submersed vegetation as habitat for invertebrates in the Hudson River estuary. *Estuaries Coasts* **2007**, *30*, 253–264. [CrossRef]
7. Dennison, W.C.; Orth, R.J.; Moore, K.A.; Stevenson, J.C.; Carter, V.; Kollar, S.; Bergstrom, P.W.; Batiuk, R.A. Assessing Water Quality with Submersed Aquatic Vegetation. *Bioscience* **1993**, *43*, 86–94. [CrossRef]
8. Ghisalberti, M.; Nepf, H.M. The limited growth of vegetated shear layers. *Water Resour. Res.* **2004**, *40*. [CrossRef]
9. Flyn, N.J.; Snook, D.L.; Wade, A.J.; Jarvie, H.P. Macrophyte and periphyton dynamics in a UK Cretaceous Chalk stream: The river Kennet, a tributary of the Thames. *Sci. Total Environ.* **2002**, *282–283*, 143–157. [CrossRef]
10. Visser, F.; Wallis, C.; Sinnott, A.M. Optical remote sensing of submerged aquatic vegetation: Opportunities for shallow clearwater streams. *Limnologica* **2013**, *43*, 388–398. [CrossRef]
11. Villa, P.; Mousivand, A.; Bresciani, M. Aquatic vegetation indices assessment through radiative transfer modeling and linear mixture simulation. *Int. J. Appl. Earth Obs. Geoinf.* **2014**, *30*, 113–127. [CrossRef]
12. Legleiter, C.J.; Roberts, D.A. A forward image model for passive optical remote sensing of river bathymetry. *Remote Sens. Environ.* **2009**, *113*, 1025–1045. [CrossRef]
13. Niroumand-Jadidi, M.; Vitti, A.; Lyzenga, D.R. Multiple Optimal Depth Predictors Analysis (MODPA) for river bathymetry: Findings from spectroradiometry, simulations, and satellite imagery. *Remote Sens. Environ.* **2018**, *218*, 132–147. [CrossRef]
14. Bailly, J.-S.; Le Coarer, Y.; Languille, P.; Stigermark, C.-J.; Allouis, T. Geostatistical estimations of bathymetric LiDAR errors on rivers. *Earth Surf. Process. Landf.* **2010**, *35*, 1199–1210. [CrossRef]
15. Mandlburger, G.; Hauer, C.; Wieser, M.; Pfeifer, N. Topo-Bathymetric LiDAR for Monitoring River Morphodynamics and Instream Habitats—A Case Study at the Pielach River. *Remote Sens.* **2015**, *7*, 6160–6195. [CrossRef]
16. Niroumand-Jadidi, M.; Vitti, A. Grain size mapping in shallow rivers using spectral information: A lab spectroradiometry perspective. *Proc. SPIE* **2017**, *10422*, 104220B.
17. Flynn, K.; Chapra, S. Remote Sensing of Submerged Aquatic Vegetation in a Shallow Non-Turbid River Using an Unmanned Aerial Vehicle. *Remote Sens.* **2014**, *6*, 12815–12836. [CrossRef]
18. Anker, Y.; Hershkovitz, Y.; Ben Dor, E.; Gasith, A. Application of aerial digital photography for macrophyte cover and composition survey in small rural streams. *River Res. Appl.* **2014**, *30*, 925–937. [CrossRef]
19. Marcus, W.A.; Fonstad, M.A. Optical remote mapping of rivers at sub-meter resolutions and watershed extents. *Earth Surf. Process. Landf.* **2008**, *33*, 4–24. [CrossRef]
20. Legleiter, C.J.; Overstreet, B.T. Mapping gravel-bed river bathymetry from space. *J. Geophys. Res.* **2012**, *117*, 82071. [CrossRef]
21. Hugue, F.; Lapointe, M.; Eaton, B.C.; Lepoutre, A. Satellite-based remote sensing of running water habitats at large riverscape scales: Tools to analyze habitat heterogeneity for river ecosystem management. *Geomorphology* **2016**, *253*, 353–369. [CrossRef]
22. Niroumand-Jadidi, M.; Vitti, A. Reconstruction of river boundaries at sub-pixel resolution: Estimation and spatial allocation of water fractions. *ISPRS Int. J. Geo-Inf.* **2017**, *6*, 383. [CrossRef]

23. Niroumand-Jadidi, M.; Vitti, A. Improving the accuracies of bathymetric models based on multiple regression for calibration (case study: Sarca River, Italy). *Proc. SPIE* **2016**, *9999*, 99990Q. [CrossRef]

24. Belward, A.S.; Skøien, J.O. Who launched what, when and why; trends in global land-cover observation capacity from civilian earth observation satellites. *ISPRS J. Photogramm. Remote Sens.* **2015**, *103*, 115–128. [CrossRef]

25. Giardino, C.; Brando, V.E.; Dekker, A.G.; Strömbeck, N.; Candiani, G. Assessment of water quality in Lake Garda (Italy) using Hyperion. *Remote Sens. Environ.* **2007**, *109*, 183–195. [CrossRef]

26. Mobley, C.D.; Sundman, L.K. *Hydrolight 5.2 Ecolight 5.2 Users' Guide*; Sequoia Scientific, Inc.: Bellevue, WA, USA, 2008.

27. DigitalGlobe. *Spectral Response for DigitalGlobe Earth Imaging Instruments*; DigitalGlobe: Westminster, CO, USA, 2013.

28. Lyzenga, D.R. Passive remote sensing techniques for mapping water depth and bottom features. *Appl. Opt.* **1978**, *17*, 379. [CrossRef]

29. Lyzenga, D.R. Remote sensing of bottom reflectance and water attenuation parameters in shallow water using aircraft and Landsat data. *Int. J. Remote Sens.* **1981**, *2*, 71–82. [CrossRef]

30. Zoffoli, M.L.; Frouin, R.; Kampel, M. Water Column Correction for Coral Reef Studies by Remote Sensing. *Sensors* **2014**, *14*, 16881–16931. [CrossRef] [PubMed]

31. Mobley, C.D. Estimation of the remote-sensing reflectance from above-surface measurements. *Appl. Opt.* **1999**, *38*, 7442. [CrossRef] [PubMed]

32. Mobley, C.D.; Sundman, L.K.; Davis, C.O.; Bowles, J.H.; Downes, T.V.; Leathers, R.A.; Montes, M.J.; Bissett, W.P.; Kohler, D.D.R.; Reid, R.P.; et al. Interpretation of hyperspectral remote-sensing imagery by spectrum matching and look-up tables. *Appl. Opt.* **2005**, *44*, 3576–3592. [CrossRef]

33. Pahlevan, N.; Schott, J.R.; Franz, B.A.; Zibordi, G.; Markham, B.; Bailey, S.; Schaaf, C.B.; Ondrusek, M.; Greb, S.; Strait, C.M. Landsat 8 remote sensing reflectance (Rrs) products: Evaluations, intercomparisons, and enhancements. *Remote Sens. Environ.* **2017**, *190*, 289–301. [CrossRef]

34. Lee, Z.; Carder, K.L.; Arnone, R.A. Deriving inherent optical properties from water color: A multiband quasi-analytical algorithm for optically deep waters. *Appl. Opt.* **2002**, *41*, 5755–5772. [CrossRef] [PubMed]

35. Maritorena, S.; Morel, A.; Gentili, B. Diffuse reflectance of oceanic shallow waters: Influence of water depth and bottom albedo. *Limnol. Oceanogr.* **1994**, *39*, 1689–1703. [CrossRef]

36. O'Neill, J.D.; Costa, M.; Sharma, T. Remote Sensing of Shallow Coastal Benthic Substrates: In situ Spectra and Mapping of Eelgrass (Zostera marina) in the Gulf Islands National Park Reserve of Canada. *Remote Sens.* **2011**, *3*, 975–1005. [CrossRef]

37. MARITORENA, S. Remote sensing of the water attenuation in coral reefs: A case study in French Polynesia. *Int. J. Remote Sens.* **1996**, *17*, 155–166. [CrossRef]

38. Mumby, P.; Edwards, A. Water Column Correction Techniques. In *Remote Sensing Handbook for Tropical Coastal Management*; UNESCO: Paris, France, 2000; pp. 121–128. ISBN 92-3-103736-6.

39. Fritz, C.; Dörnhöfer, K.; Schneider, T.; Geist, J.; Oppelt, N. Mapping Submerged Aquatic Vegetation Using RapidEye Satellite Data: The Example of Lake Kummerow (Germany). *Water* **2017**, *9*, 510. [CrossRef]

40. Stumpf, R.P.; Holderied, K.; Sinclair, M. Determination of water depth with high-resolution satellite imagery over variable bottom types. *Limnol. Oceanogr.* **2003**, *48*, 547–556. [CrossRef]

41. Niroumand-Jadidi, M.; Vitti, A. Optimal band ratio analysis of WorldView-3 imagery for bathymetry of shallow rivers (case study: Sarca River, Italy). Available online: https://www.researchgate.net/publication/304343285_OPTIMAL_BAND_RATIO_ANALYSIS_OF_WORLDVIEW-3_IMAGERY_FOR_BATHYMETRY_OF_SHALLOW_RIVERS_CASE_STUDY_SARCA_RIVER_ITALY (accessed on 28 January 2019).

42. Legleiter, C.J.; Roberts, D.A.; Lawrence, R.L. Spectrally based remote sensing of river bathymetry. *Earth Surf. Process. Landf.* **2009**, *34*, 1039–1059. [CrossRef]

43. Flener, C.; Lotsari, E.; Alho, P.; Käyhkö, J. Comparison of empirical and theoretical remote sensing based bathymetry models in river environments. *River Res. Appl.* **2012**, *28*, 118–133. [CrossRef]

44. Flener, C. Estimating Deep Water Radiance in Shallow Water: Adapting Optical Bathymetry Modelling to Shallow River Environments. *Boreal Environ. Res.* **2013**, *18*, 488–502.

45. Hartigan, J.A.; Wong, M.A. Algorithm AS 136: A K-Means Clustering Algorithm. *Appl. Stat.* **1979**, *28*, 100. [CrossRef]

46. Jensen, J.R. *Introductory Digital Image Processing: A Remote Sensing Perspective*; Prentice Hall: Upper Saddle River, NJ, USA, 2005; ISBN 0131453610.

47. Cho, H.J.; Mishra, D.; Wood, J. Remote Sensing of Submerged Aquatic Vegetation. In *Remote Sensing—Applications*; InTech: Rijeka, Croatia, 2012; ISBN 9789535106517.

48. Hedley, J.D.; Roelfsema, C.M.; Phinn, S.R.; Mumby, P.J.; Hedley, J.D.; Roelfsema, C.M.; Phinn, S.R.; Mumby, P.J. Environmental and Sensor Limitations in Optical Remote Sensing of Coral Reefs: Implications for Monitoring and Sensor Design. *Remote Sens.* **2012**, *4*, 271–302. [CrossRef]

49. Hurcom, S.J.; Harrison, A.R. The NDVI and spectral decomposition for semi-arid vegetation abundance estimation. *Int. J. Remote Sens.* **1998**, *19*, 3109–3125. [CrossRef]

50. Elmore, A.J.; Mustard, J.F.; Manning, S.J.; Lobell, D.B. Quantifying Vegetation Change in Semiarid Environments: Precision and Accuracy of Spectral Mixture Analysis and the Normalized Difference Vegetation Index. *Remote Sens. Environ.* **2000**, *73*, 87–102. [CrossRef]

51. Mobley, C.D. *Light and Water: Radiative Transfer in Natural Waters*; Academic Press: San Diego, CA, USA, 1994; ISBN 9780125027502.

52. Gordon, H.R.; Ding, K. Self-shading of in-water optical instruments. *Limnol. Oceanogr.* **1992**, *37*, 491–500. [CrossRef]

53. Niroumand-Jadidi, M.; Bovolo, F.; Vitti, A.; Bruzzone, L. A novel approach for bathymetry of shallow rivers based on spectral magnitude and shape predictors using stepwise regression. In *Image and Signal Processing for Remote Sensing XXIV*; Bruzzone, L., Bovolo, F., Benediktsson, J.A., Eds.; SPIE Remote Sensing: Berlin, Germany, 2018; Volume 10789, p. 23.

54. Legleiter, C.J.; Roberts, D.A. Effects of channel morphology and sensor spatial resolution on image-derived depth estimates. *Remote Sens. Environ.* **2005**, *95*, 231–247. [CrossRef]

55. Lane, C.; Liu, H.; Autrey, B.; Anenkhonov, O.; Chepinoga, V.; Wu, Q.; Lane, C.R.; Liu, H.; Autrey, B.C.; Anenkhonov, O.A.; et al. Improved Wetland Classification Using Eight-Band High Resolution Satellite Imagery and a Hybrid Approach. *Remote Sens.* **2014**, *6*, 12187–12216. [CrossRef]

56. Whiteside, T.G.; Bartolo, R.E. Mapping Aquatic Vegetation in a Tropical Wetland Using High Spatial Resolution Multispectral Satellite Imagery. *Remote Sens.* **2015**, *7*, 11664–11694. [CrossRef]

57. Berk, A.; Anderson, G.P.; Acharya, P.K.; Bernstein, L.S.; Muratov, L.; Lee, J.; Fox, M.; Adler-Golden, S.M.; Chetwynd, J.H., Jr.; Hoke, M.L.; et al. *MODTRAN5: 2006 Update*; Shen, S.S., Lewis, P.E., Eds.; International Society for Optics and Photonics: Orlando, FL, USA, 2006; Volume 6233, p. 62331F.

58. Pahlevan, N.; Lee, Z.; Wei, J.; Schaaf, C.B.; Schott, J.R.; Berk, A. On-orbit radiometric characterization of OLI (Landsat-8) for applications in aquatic remote sensing. *Remote Sens. Environ.* **2014**, *154*, 272–284. [CrossRef]

59. Carbonneau, P.E.; Lane, S.N.; Bergeron, N. Feature based image processing methods applied to bathymetric measurements from airborne remote sensing in fluvial environments. *Earth Surf. Process. Landf.* **2006**, *31*, 1413–1423. [CrossRef]

60. Overstreet, B.T.; Legleiter, C.J. Removing sun glint from optical remote sensing images of shallow rivers. *Earth Surf. Process. Landf.* **2017**, *42*, 318–333. [CrossRef]

61. Fonstad, M.A.; Marcus, W.A. Remote sensing of stream depths with hydraulically assisted bathymetry (HAB) models. *Geomorphology* **2005**, *72*, 320–339. [CrossRef]

MDPI

St. Alban-Anlage 66

4052 Basel

Switzerland

Tel. +41 61 683 77 34

Fax +41 61 302 89 18

www.mdpi.com

Remote Sensing Editorial Office

E-mail: remotesensing@mdpi.com

www.mdpi.com/journal/remotesensing

www.ingramcontent.com/pod-product-compliance
Lightning Source LLC
Chambersburg PA
CBHW051911210326
41597CB00033B/6109